Lecture Notes in Mathematics

Edited by A. Dold and B. Eckmann
Series: Forschungsinstitut für Mathematik, ETH Zürich

433

William G. Faris

T0219863

Self-Adjoint Operators

Springer-Verlag
Berlin · Heidelberg · New York 1975

William G. Faris
Battelle Institute
Advanced Studies Center
1227 Carouge-Genève/Suisse

Present Address:
Dept. of Mathematics
University of Arizona
Tucson, AZ 85721/USA

Library of Congress Cataloging in Publication Data

Faris, William G 1939-
Self-adjoint operators.

(Lecture notes in mathematics ; 433)
Includes bibliographical references and index.
1. Selfadjoint operators. 2. Perturbation (Mathematics) 3. Quantum theory. I. Title. II. Series: Lecture notes in mathematics (Berlin) ; 433.
QA3.L28 no. 433 [QA329.2] 510'.8s [515'.72] 74-32497

AMS Subject Classifications (1970): 47-02, 47A55, 81A09, 81A10

ISBN 3-540-07030-3 Springer-Verlag Berlin · Heidelberg · New York
ISBN 0-387-07030-3 Springer-Verlag New York · Heidelberg · Berlin

Offsetdruck: Julius Beltz, Hemsbach/Bergstr.

PREFACE

These lectures are primarily concerned with the problem of when the sum of two unbounded self-adjoint operators is self-adjoint. The sum fails to be self-adjoint when there is an ambiguity about the choice of boundary conditions. There is then also an ambiguity in the definition of functions of the sum.

This problem is fundamental to quantum mechanics. Kinetic energy and potential energy are self-adjoint operators, and functions of either of these may be computed explicitly. The sum is the total energy, and the main task of quantum mechanics is to compute functions of it. Thus the question of self-adjointness is the question of whether this task is meaningful. If it is, then among other things the dynamics of the system is determined for all states and all times.

There are obvious series expansions for certain important functions, such as the resolvent. The point is that these series may not converge. In order to demonstrate self-adjointness it is necessary to develop alternative methods for approximating functions of the sum.

In the language of physics, the basic question is whether the force laws determine the motion. The models currently used contain elements which could lead to ambiguity. For example, the charge on an electron is considered to be squeezed down to one point. As a consequence its potential energy is unbounded. What happens when two electrons occupy the same point? Their interaction energy is infinite, and it is not clear how they will move.

Actually, according to quantum mechanics it is improbable that two electrons will be at or even near the same point. But there are similar difficulties in the description of light by quantum field theory, and these are not so easy to resolve. The question of whether this theory gives unambigous predictions to an arbitrarily high degree of accuracy remains open. So it is worth while to examine the mechanism for determining the dynamics in better understood situations.

The justification for these lectures is that there has been progress on the self-adjointness problem since the publication of Kato's book, Perturbation Theory for Linear Operators (1966). This progress has been stimulated largely by developments in quantum field theory. While the results which depend on a series expansion have not been significantly improved, those which exploit positivity are the heart of the recent developments. This is perhaps because for many physical systems one expects the total energy to be bounded below.

While Kato's book is the basic reference for linear perturbation theory, there are other books with additional material on the quantum mechanical applications. Those by Hellwig (1967) and Glazman (1965) approach the subject with partial differential equation techniques. More recently new progress has been made on the determination of the spectra of quantum mechanical operators, and this is described in the books of Simon (1971a), Schechter (1971), and Jörgens and Weidmann (1973).

Recently a memoir of Chernoff (1974) has appeared. It contains a valuable discussion of the addition problem which complements the present treatment.

These lectures begin with a review of standard material on addition of self-adjoint operators. The second part is a selection from the more recent developments. This includes a section on properties of eigenvalues, including uniqueness of the ground state. The third part is devoted to the classification of extensions of a Hermitian operator. This is standard material; however it is there to illustrate what kind of ambiguity is possible when self-adjointness fails. (After all, to understand the force of a theorem you have to be able to imagine a situation where its conclusion fails.) The final part is a brief account of how self-adjointness is related to the determination of a measure by its moments.

The core of the lectures is the first two parts. The theme is the interplay between the two aspects of a quantum mechanical observable, as an operator and as a quadratic form. In order to be able to take functions of an observable it is necessary for it to be a self-adjoint operator. But to add observables it is most natural to add quadratic forms (that is, to add expectation values).

Throughout, the theory is illustrated by the case of the Schrödinger equation for a non-relativistic particle in a given potential. The emphasis is on results obtained by operator theory rather than by partial differential equation methods. (There is almost no discussion of the case of ordinary differential equations, but this subject has recently been surveyed by Devinatz (1973).) There is a brief description of the applications to quantum field theory.

These lectures are intended to be an introduction to one topic in operator theory. They are not a complete treatment even of this topic, but should be regarded as an invitation to the research literature. In order to follow them it should be sufficient to know real analysis and have some acquaintance with Hilbert space. The spectral theorem is stated but not proved.

The lectures were first given (in a somewhat different form) at the Eidgenössische Technische Hochschule in Zürich during the spring of 1971. I wish to thank Professor Barry Simon for references to the literature, and Dr. Jean-Pierre Eckmann, Dr. Charles Stuart, and Dr. Lawrence Thomas for reading the manuscript. I am especially grateful to Miss Edeltraud Russo for her excellent secretarial work.

W. Faris

Geneva
August, 1974

CONTENTS

Part I : FORMS AND OPERATORS

§1 OPERATORS

In this section we review the basic facts about Hilbert space and about self-adjoint operators acting in Hilbert space. The goal is the statement of the spectral theorem.

A Hilbert space is a vector space with a certain type of form defined on it, so we begin with forms.

Let E be a complex vector space. A sesquilinear form S is a mapping which assigns to each ordered pair f , g of elements of E a complex number $S(f,g)$ in such a way that the correspondence is conjugate linear in f and linear in g . (Thus $S(af,g) = a^*S(f,g)$, where a^* is the complex conjugate of the complex number a .)

There is a quadratic form associated with every sesquilinear form, the one which assigns to every f in E the complex number $S(f,f)$. The sesquilinear form may be recovered from the quadratic form by the polarization identity

$$4S(f,g) = S(g+f,g+f) + iS(g+if,g+if) - S(g-f,g-f) - iS(g-if,g-if) .$$

Thus we sometimes also refer to the sesquilinear form as being defined on E .

The sesquilinear form S is said to be Hermitian if $S(f,g) = S(g,f)^*$ for all f and g in E .

Proposition 1.1 . A sesquilinear form is Hermitian if and only if the associated quadratic form is real.

Proof: If $S(f,f)$ is real, then $S(f,g)$ and $S(g,f)^*$ have the same quadratic form $S(f,f) = S(f,f)^*$.

A sesquilinear form is called positive or strictly positive if its quadratic form is positive or strictly positive (except at zero).

An inner product is a strictly positive sesquilinear form. It is customary to denote an inner product by $\langle f,g\rangle$ and the associated quadratic form by $\|f\|^2$ (the square of the norm $\|f\|$).

An inner product satisfies the Schwarz inequality $|\langle f,g\rangle| \leqslant \|f\|\,\|g\|$. It follows that the norm $\|g\|$ can be computed as the supremum of the $|\langle f,g\rangle|$ with $\|f\| \leqslant 1$.

Another consequence of the Schwarz inequality is that if g is fixed, $f \mapsto \langle f,g\rangle$ is a continuous conjugate linear functional on E . This correspondence between elements of E and functionals is injective, since if $\langle f,g\rangle = 0$ for all f in E , in particular $\langle g,g\rangle = 0$, and so $g = 0$. (Similarly, if f is fixed, $g \mapsto \langle f,g\rangle$ is a continuous linear functional on E .)

Elements f and g of an inner product space are called orthogonal if their inner product $\langle f,g\rangle = 0$. If f and g are orthogonal we write $f \perp g$.

Let M be a linear subspace of the inner product space E . Define $M^\perp$ to be the set of f in E such that $f \perp g$ for all g in M . The following assertions are evident:

(i) $M^\perp$ is a closed subspace of E

(ii) $\overline{M}^\perp = M^\perp$

(iii) If $M \subset N$ then $N^\perp \subset M^\perp$.

Notice also that if M is dense in E , then from (ii) we have $M^{\perp} = E^{\perp} = \{0\}$.

If f is in E and $M \subset E$, then g in M is called the <u>orthogonal projection</u> of f on M if $f = g + h$ where h is in $M^{\perp}$.

A (complex) <u>Hilbert space</u> is a complex vector space with an inner product, which is complete in the associated norm.

<u>Projection Theorem 1.2</u> . Let H be a Hilbert space and $M \subset H$ be a closed linear subspace. Then every element of H has an orthogonal projection on M . That is, $H = M \oplus M^{\perp}$.

This fundamental existence theorem is stated without proof.

Let M be a linear subspace of H . Then by the projection theorem applied to $\overline{M}$, $H = \overline{M} \oplus M^{\perp}$. On the other hand, by the projection theorem applied to $M^{\perp}$, $H = M^{\perp\perp} \oplus M^{\perp}$. Since $\overline{M} \subset M^{\perp\perp}$, it follows that

(iv) $\overline{M} = M^{\perp\perp}$.

It also follows from the projection theorem that if M is a linear subspace of the Hilbert space H , then $\overline{M} = H$ if and only if $M^{\perp} = \{0\}$. In fact, if $M^{\perp} = \{0\}$, then by (iv) $\overline{M} = \{0\}^{\perp} = H$.

<u>Riesz Representation Theorem 1.3</u> . Let H be a Hilbert space. Then any continuous conjugate linear functional on H is of the form $f \mapsto \langle f,g \rangle$ for some element g of H . (In other words, the correspondence between elements and functionals is surjective.)

<u>Proof</u>: The kernel of the functional L is a closed linear subspace M of H . If $M = H$ we take $g = 0$. Otherwise $M^{\perp}$ contains an

element $h \neq 0$ (by the projection theorem). We take $g = ch$, where c is a complex number chosen so that $(Lg) = \langle g,g\rangle$. If f is in H, then $f - (Lf)^*(Lg)^{*-1}g$ is in M, so
$\langle f,g\rangle = Lf\,(Lg)^{-1}\langle g,g\rangle = Lf$.

Of course there is an analogous representation theorem for continuous linear functionals on H.

Let H and K be Hilbert spaces. A linear transformation $U : H \rightarrow K$ is said to be <u>unitary</u> if it is bijective and preserves inner products. In other words, a unitary transformation is an isomorphism of Hilbert spaces.

Next we turn to the objects of main interest, operators. Let H be a Hilbert space. An <u>operator</u> A is a linear transformation from some linear subspace $\mathcal{D}(A)$ of H to H.

A <u>graph</u> is a linear subspace A of $H \oplus H$. The domain $\mathcal{D}(A)$ of a graph is the space of f with $f \oplus g$ in A. A graph A defines an operator if whenever $f \oplus g$ belongs to A and $f = 0$, then $g = 0$.

Let $\tau : H \oplus H \rightarrow H \oplus H$ be defined by $\tau(f \oplus g) = (-g \oplus f)$. Then τ is unitary and $\tau^2 = -1$.

* If A is a graph, its <u>adjoint</u> A^* is the graph defined by $A^* = (\tau A)^{\perp} = \tau(A^{\perp})$.

From the properties of the orthogonal complement we conclude that

(i) A^* is closed (as a subspace of $H \oplus H$).

(ii) $\bar{A}^* = A^*$

(iii) If $A \subset B$ then $B^* \subset A^*$

(iv) $A^{**} = \bar{A}$.

Here $\overline{A}$ is the closure of the graph A in $\mathcal{H} \oplus \mathcal{H}$.

<u>Proposition 1.4</u> . Let A be a graph. Then A^* is an operator if and only if A is densely defined.

<u>Proof</u>: $A^* \cap (\{0\} \oplus \mathcal{H}) = \{0 \oplus 0\}$ if and only if $A^{\perp} \cap (\mathcal{H} \oplus \{0\}) = \{0 \oplus 0\}$ if and only if $\mathcal{D}(A)^{\perp} = \{0\}$ if and only if $\overline{\mathcal{D}(A)} = \mathcal{H}$.

<u>Corollary</u> . Let A be a graph. Then $\overline{A}$ is an operator if and only if A^* is densely defined.

<u>Proof</u>: $\overline{A} = (A^*)^*$.

If $\overline{A}$ is an operator then A is said to be <u>closable</u>. Notice that A is densely defined and closable if and only if A^* is an operator which is densely defined. Thus the class of densely defined closable operators is well-behaved under the operation of taking the adjoint.

<u>Proposition 1.5</u> . Let A be a densely defined operator. Then A^* is an operator. Its domain $\mathcal{D}(A^*) = \{g$ in $\mathcal{H} : f \mapsto \langle Af,g\rangle$ is continuous in the norm of $\mathcal{H}\}$ and $\langle f,A^*g\rangle = \langle Af,g\rangle$ for f in $\mathcal{D}(A)$ and g in $\mathcal{D}(A^*)$.

<u>Proof</u>: $A^*g = h$ if and only if $g \oplus h \perp \tau(f \oplus Af)$ if and only if $g \oplus h \perp (-Af \oplus f)$ if and only if $\langle Af,g\rangle = \langle f,h\rangle$ (for all f in $\mathcal{D}(A)$).

<u>Proposition 1.6</u> . Let A be a densely defined operator. Then the kernel of A^* is $(\text{range } A)^{\perp}$.

<u>Proof</u>: The element g is in $\mathcal{D}(A^*)$ with $A^*g = 0$ precisely when $\langle Af,g\rangle = 0$ for all f in $\mathcal{D}(A)$.

Corollary . Let A be a densely defined closed operator. Assume that A is injective and A^{-1} is continuous. Then $\mathcal{H} = (\text{range } A) \oplus \text{kernel } A^*$.

Proof: The map $Af \mapsto f \oplus Af$ from range A to the graph of A is the same as the map $h \mapsto A^{-1}h \oplus h$, which is continuous. Since the graph of A is closed, so is range A .

An operator A is said to be *Hermitian* if A is densely defined and $A \subset A^*$. It is said to be *self-adjoint* if $A = A^*$.

Notice that the operator A satisfies $A \subset A^*$ if and only if $\langle Af,g\rangle = \langle f,Ag\rangle$ for all f and g in $\mathcal{D}(A)$. Since $\langle Af,g\rangle = \langle g,Af\rangle^*$, this equation says precisely that the form $\langle f,Ag\rangle$ is Hermitian.

If A is a graph, its inverse A^{-1} is the graph consisting of all $g \oplus f$ such that $f \oplus g$ is in A . Clearly $A^{-1*} = A^{*-1}$.

Theorem 1.7 . Let A be an operator such that $\langle Af,g\rangle = \langle f,Ag\rangle$ for all f and g in $\mathcal{D}(A)$. If there exists a complex number z such that range $A - z = \mathcal{H}$ and range $A - z^* = \mathcal{H}$, then A is self-adjoint.

Proof: Let A^* be the graph which is the adjoint of A . Then $A \subset A^*$, so $A-z \subset A^*-z$, $(A-z)^{-1} \subset (A^*-z)^{-1}$. Since range $A - z^* = \mathcal{H}$, $\mathcal{D}((A-z^*)^{-1}) = \mathcal{H}$. Thus $(A-z^*)^{-1*} = (A^*-z)^{-1}$ is an operator. Since range $A - z = \mathcal{H}$, $\mathcal{D}((A-z)^{-1}) = \mathcal{H}$. Thus $(A-z)^{-1} \subset (A^*-z)^{-1}$ implies $(A-z)^{-1} = (A^*-z)^{-1}$, $A = A^*$.

The point of this theorem is that the property of A being self-adjoint is related to the existence of the *resolvent* operator $(A-z)^{-1} : \mathcal{H} \to \mathcal{H}$ (for a pair of complex conjugate z).

Let μ be a positive measure in the measure space M. Then $L^2(M,\mu)$ is a Hilbert space.

Let α be a measurable complex function on M. The corresponding <u>multiplication operator</u> $\mathrm{mult}(\alpha)$ acting in L^2 is defined by $g \mapsto \alpha g$ whenever g and αg are in L^2.

If α is real then $\mathrm{mult}(\alpha)$ is self-adjoint (by Theorem 1.7).

<u>Spectral Theorem 1.8</u>. Let A be a self-adjoint operator acting in the Hilbert space $\mathcal{H}$. Then A is isomorphic to a multiplication operator. That is, there exists a Hilbert space $L^2(M,\mu)$ of square integrable functions, a unitary operator $U : \mathcal{H} \to L^2(M,\mu)$, and a real measurable function α on M such that $UA = \mathrm{mult}(\alpha)U$.

The spectral theorem will not be proved here.

The point of this theorem is that any question about a single self-adjoint operator is a question about a function. Thus in the following we usually will regard assertions about a single self-adjoint operator as obvious and not needing proof. The difficulties begin with two (non-commuting) self-adjoint operators.

Let A be a self-adjoint operator acting in $\mathcal{H}$. Then A is isomorphic to multiplication by some real function α. Let ϕ be a complex Borel measurable function on the reals. Then $\phi(A)$ is defined as the operator which is isomorphic (by the same unitary operator) to multiplication by $\phi(\alpha)$. It may be proved that this definition does not depend on the choice of the isomorphism.

The most important of these functions ϕ are the inverse, complex exponential, and indicator functions. These give rise to the resolvent, unitary group, and spectral projections of the self-adjoint operator.

There is a set of complex numbers z such that the resolvent $(A-z)^{-1}: H \to H$ is continuous. Its complement is called the <u>spectrum</u> of A. The spectrum of a self-adjoint operator A is a closed subset of the real axis. If A is isomorphic to multiplication by a real function α, the spectrum of A is the essential range of α.

Multiplication by a function of absolute value one is a unitary operator. Thus the complex exponentials of a self-adjoint operator form a one-parameter <u>unitary group</u>. There is a close relation between the resolvent and the unitary group. It is given by

$$(A-z)^{-1}f = i\int_0^{\pm\infty} \exp(itz)\exp(-itA)f\,dt$$

for $\pm\mathrm{Im}z > 0$ and the inversion formula $\exp(-itA)f = \lim_{n\to\infty}(1+\frac{it}{n}A)^{-n}f$.

Let S be a Borel subset of the real line and write 1_S for the indicator function of S (which is 1 on S and 0 on its complement). The <u>spectral projection</u> of A corresponding to S is $1_S(A)$.

The resolvent may also be used to approximate spectral projections. For $\varepsilon > 0$ let $\delta(\varepsilon;s) = \pi^{-1}(\varepsilon/(s^2+\varepsilon^2))$. As $\varepsilon \to 0$, the functions $\int_a^b \delta(\varepsilon;s-x)dx$ converge to 1 if s is in (a,b) and to 0 if s is in the complement of $[a,b]$. Thus the operators $\delta(\varepsilon;A-x)$ are relevant to computation of certain spectral projections. But these operators may be expressed in terms of resolvents by $\delta(\varepsilon;A-x) = (2\pi i)^{-1}[(A-z)^{-1} - (A-z^*)^{-1}]$, where $z = x + i\varepsilon$.

When an operator A is shown to be self-adjoint, the proof usually leads to an approximation procedure for certain functions of the operator. These are most often the resolvent $(A-z)^{-1}$ for z

far from the real axis or the exponential $\exp(itA)$ for t not too large. These functions give a certain amount of useful information about asymptotic properties of the spectrum, but to find detailed spectral properties requires knowledge of the resolvent near the real axis or of the large time limit of the exponential. This is a more delicate matter, and is beyond the scope of these lectures. There is a huge amount of information on the determination of the spectrum, however. In fact, it might be said to be the main enterprise of theoretical physics.

EXAMPLES

An example of a unitary operator from one Hilbert space to another is the Fourier transform $F : L^2(\mathbb{R}^n, dx) \rightarrow L^2(\mathbb{R}^n, dk/(2\pi)^n)$. It is determined on a dense set by the formula $Fg(k) = \hat{g}(k) = \int \exp(-ikx)g(x)dx$. Its inverse is determined by $h(x) = \int \exp(ikx)\hat{h}(k)dk/(2\pi)^n$. The importance of the Fourier transform is that it gives the spectral representation of translation invariant operators.

Consider first the one-dimensional case. Let $H = L^2(\mathbb{R}, dx)$ and let $A = -\frac{d^2}{dx^2}$, defined on the subspace of f in L^2 such that f and f' are absolutely continuous and f' and f'' are in L^2 . Then A is self-adjoint. In fact $(A+c)^{-1}$ is given by

$$(A+c)^{-1}h(x) = \tfrac{1}{2}c^{-\frac{1}{2}} \int \exp(-c^{\frac{1}{2}}|x-y|)h(y)dy$$

except when $c \leqslant 0$. Thus the spectrum of A is the positive real axis.

According to the spectral theorem, A is isomorphic to a multiplication operator. In this case the isomorphism is given by the Fourier transform, and A is isomorphic to multiplication by k^2 .

In n dimensions the analogous operator is $A = -\Delta$. It is perhaps simplest to define this operator by means of the Fourier transform, so that by definition it is isomorphic to multiplication by k^2 .

NOTES

The most succint self-contained treatment of Hilbert space (including the spectral theorem) is §5 of the notes by Nelson (1969). The spectral theorem is also proved in Dunford and Schwartz (1963).

A good treatment of Fourier transforms is Stein and Weiss (1971).

§2 FORMS

We now turn to the study of sesquilinear forms defined on a linear subspace of a Hilbert space. There will always be at least two forms under consideration, the given sesquilinear form and the inner product of the Hilbert space. (In quantum mechanics the corresponding quadratic forms give the expected energy and total probability.) The relation between these two forms is the central theme of this section.

We begin with an important example. Let A be a positive self-adjoint operator acting in the Hilbert space $\mathcal{H}$. The form domain $\mathcal{Q}(A)$ of A is the linear subspace of all f in $\mathcal{H}$ such that $\|A^{\frac{1}{2}}f\|$ is finite. The form of A is defined on $\mathcal{Q}(A)$ by $S(f,g) = \langle A^{\frac{1}{2}}f, A^{\frac{1}{2}}g\rangle$.

Notice that the operator domain $\mathcal{D}(A)$ of a self-adjoint operator A consists of the f in $\mathcal{H}$ such that $\|Af\|$ is finite. Since $\|A^{\frac{1}{2}}f\|^2 \leqslant \|f\|\,\|Af\|$, the form domain of a positive self-adjoint operator is larger than its operator domain.

It is illuminating to see what these objects look like in a spectral representation. Let U be a unitary correspondence between $\mathcal{H}$ and $L^2(M,\mu)$ which establishes an isomorphism of $A \geqslant 0$ with multiplication by $\alpha \geqslant 0$. Then $\mathcal{Q}(A)$ corresponds to $L^2(M,(1+\alpha)\mu)$ and the form of A is given by $S(f,g) = \int (Uf)^* \alpha (Ug)\,d\mu$. Notice that $\mathcal{D}(A)$ corresponds to $L^2(M,(1+\alpha^2)d\mu)$.

Let S be a sesquilinear form on a linear subspace $\mathcal{E}$ of the Hilbert space $\mathcal{H}$. Assume that $\mathcal{E}$ is dense in $\mathcal{H}$. The associated operator A is defined on $\mathcal{D}(A) = \{g \text{ in } \mathcal{E} : f \mapsto S(f,g) \text{ is continuous in the norm of } \mathcal{H}\}$ by $S(f,g) = \langle f, Ag\rangle$.

In the above example, that of the form $S(f,g) = \langle A^{\frac{1}{2}}f, A^{\frac{1}{2}}g\rangle$ of a positive self-adjoint operator, the associated operator is just the original operator $A : \mathcal{D}(A) \to H$.

It is possible to characterize the forms of positive self-adjoint operators. For this we need one more definition.

Let S be a positive form defined on the linear subspace $\mathcal{Q}$ of the Hilbert space $\mathcal{H}$. Notice that $\langle f,g\rangle_{\mathcal{Q}} = S(f,g) + \langle f,g\rangle$ is an inner product on $\mathcal{Q}$. We say S is <u>closed</u> if $\mathcal{Q}$ is a Hilbert space with this inner product.

The form of a positive self-adjoint operator is closed.

The next theorem states that there is a bijective correspondence between positive self-adjoint operators and densely defined, closed, positive forms.

<u>Form Representation Theorem 2.1</u> . Let $\mathcal{H}$ be a Hilbert space and $\mathcal{Q}$ be a dense linear subspace. Let S be a closed positive form on $\mathcal{Q}$. Then the associated operator A is a positive self-adjoint operator and S is the form of A .

<u>Proof</u>: In order to show that A is self-adjoint it is sufficient to show that range $(A+1) = \mathcal{H}$. Let h be an element of $\mathcal{H}$. Then the functional $f \mapsto \langle f,h\rangle$ is a continuous conjugate linear functional on the Hilbert space $\mathcal{Q}$. Hence by the Riesz representation theorem there is an element g in $\mathcal{Q}$ such that
$\langle f,h\rangle = \langle f,g\rangle_{\mathcal{Q}} = S(f,g) + \langle f,g\rangle$. Thus g is in $\mathcal{D}(A)$ and $(A+1)g = h$.

Next we show that $\mathcal{D}(A)$ is dense in $\mathcal{Q}$. Let h in $\mathcal{Q}$ be orthogonal to $\mathcal{D}(A)$ in the inner product of $\mathcal{Q}$. Since h is in $\mathcal{H}$, we may write $h = (A+1)g$, where g is in $\mathcal{D}(A)$. But then $\langle h,h\rangle = \langle h,(A+1)g\rangle = \langle h,g\rangle_{\mathcal{Q}} = 0$, so $h = 0$.

The inner products on $\mathcal{Q}$ and on the form domain $\mathcal{Q}(A)$ of A coincide on $\mathcal{D}(A)$, and $\mathcal{D}(A)$ is dense in both spaces. So the inner products coincide on $\mathcal{Q} = \mathcal{Q}(A)$ and S is the form of A .

Remark . It is possible to think of this construction in another way. Let $\mathcal{Q}^*$ be the conjugate dual space of $\mathcal{Q}$ (the space of continuous conjugate linear functionals on $\mathcal{Q}$). If h is an element of $\mathcal{H}$, we may identify it with the element $f \mapsto \langle f,h\rangle$ of $\mathcal{Q}^*$. Thus we may regard $\mathcal{Q} \subset \mathcal{H} \subset \mathcal{Q}^*$ and it is consistent to write $\langle f,v\rangle$ for the value of v in $\mathcal{Q}^*$ on f in $\mathcal{Q}$.

If g is in $\mathcal{Q}$, define Ag in $\mathcal{Q}^*$ by $\langle f,Ag\rangle = S(f,g)$. Then the operator A acting in $\mathcal{H}$ associated with S is just the restriction of $A : \mathcal{Q} \to \mathcal{Q}^*$ to $\mathcal{D}(A) = \{g \text{ in } \mathcal{Q} : Ag \text{ is in } \mathcal{H}\}$.

The Riesz representation theorem says that $A + 1 : \mathcal{Q} \to \mathcal{Q}^*$ is surjective, and so the restriction $A + 1 : \mathcal{D}(A) \to \mathcal{H}$ is also surjective. This is what is needed for self-adjointness.

The conjugate dual space $\mathcal{Q}^*$ is a Hilbert space with the norm $\|v\|_{\mathcal{Q}^*} = \sup\{|\langle f,v\rangle| : \|f\|_{\mathcal{Q}} \leq 1\}$. Since $\langle f,(A+1)g\rangle = \langle f,g\rangle_{\mathcal{Q}}$, we have $\|(A+1)g\|_{\mathcal{Q}^*} = \|g\|_{\mathcal{Q}}$. Hence the operator $A + 1 : \mathcal{Q} \to \mathcal{Q}^*$ is unitary. The inner product on $\mathcal{Q}^*$ may be expressed as

$$\langle u,v\rangle_{\mathcal{Q}^*} = \langle (A+1)^{-1}u,(A+1)^{-1}v\rangle_{\mathcal{Q}} = \langle (A+1)^{-1}u,v\rangle .$$

Next notice that $\mathcal{H}$ is dense in $\mathcal{Q}^*$. In fact, if f is in $\mathcal{Q}$ with $\langle f,h\rangle = 0$ for all h in $\mathcal{H}$, then $\langle f,f\rangle = 0$ and so $f = 0$.

The fact that $\mathcal{D}(A)$ is dense in $\mathcal{Q}$ may now be seen as follows. Since $A + 1 : \mathcal{Q} \rightarrow \mathcal{Q}^*$ is unitary, this is equivalent to $(A+1)\mathcal{D}(A) = \mathcal{H}$ being dense in $(A+1)\mathcal{Q} = \mathcal{Q}^*$, which we have just seen to be true.

It is often easy to verify that a form is densely defined and positive. To show that it is closed may be more difficult. Is it possible to produce closed forms at will by a process of completion? The answer is negative.

Let S be a positive form on a linear subspace $\mathcal{D}$ of the Hilbert space $\mathcal{H}$. Then $\langle f,g\rangle_{\mathcal{D}} = S(f,g) + \langle f,g\rangle$ is an inner product on $\mathcal{D}$. The inclusion map of $\mathcal{D}$ into $\mathcal{H}$ is continuous. Let $\mathcal{Q}$ be the Hilbert space which is the completion of $\mathcal{D}$. If the extension by continuity of the inclusion to a map from $\mathcal{Q}$ to $\mathcal{H}$ is injective, then $\mathcal{Q}$ may also be regarded as contained in $\mathcal{H}$.

Let $\mathcal{H}$ be a Hilbert space. Let S be a positive form on a linear subspace $\mathcal{D}$ of $\mathcal{H}$. Let $\mathcal{Q}$ be the completion of $\mathcal{D}$ in the inner product $\langle f,g\rangle_{\mathcal{D}} = S(f,g) + \langle f,g\rangle$. Then S is said to be closable if $\mathcal{Q}$ is contained in $\mathcal{H}$. In that case the closure $\overline{S}$ of S is the positive form defined on $\mathcal{Q}$ by $\langle f,g\rangle_{\mathcal{Q}} = \overline{S}(f,g) + \langle f,g\rangle$.

Let A be a positive self-adjoint operator and consider the form $\langle f,Ag\rangle$ for f and g in $\mathcal{D}(A)$. This form is closable (but not necessarily closed). In fact, since $\mathcal{D}(A)$ is dense in $\mathcal{Q}(A)$, its closure is the form of A.

EXAMPLES

Let $H = L^2(\mathbb{R},dx)$ and $A = -\frac{d^2}{dx^2}$. Since A is isomorphic to multiplication by k^2 in the Fourier transform representation, the form of A is given by $S(f,g) = \int \hat{f}^* k^2 \hat{g}\, dk/2\pi = \langle k\hat{f}, k\hat{g}\rangle = \langle \frac{d}{dx}f, \frac{d}{dx}g\rangle$.

If g is in the form domain Q of A, then its Fourier transform $\hat{g}$ is integrable: $\|\hat{g}\|_1 \leq \|(k^2+1)^{-\frac{1}{2}}\|_2 \|(k^2+1)^{\frac{1}{2}}\hat{g}\|_2 < \infty$. Hence g is continuous and bounded.

The dual space Q^* of Q is the completion of H in the inner product $\langle u,v\rangle_{Q^*} = \langle u,(A+1)^{-1}v\rangle = \frac{1}{2}\iint u(x)^* \exp(-|x-y|)v(y)\,dx\,dy$. Thus the elements of Q^* need not even be functions.

Next we give an example of a densely defined positive form which is not closable. Let $H = L^2(\mathbb{R},dx)$ and let D be the form domain of $-\frac{d^2}{dx^2}$. Define $S(f,g) = f(0)^* g(0) = \int \hat{f}(k)^* \frac{dk}{2\pi} \int \hat{g}(k) \frac{dk}{2\pi}$ for f and g in D.

The completion of D in the norm $|f(0)|^2 + \int |f(x)|^2 dx$ is not contained in H. In fact, for any complex number z there is a sequence f_n in D such that $f_n(0) = z$ and $f_n \to 0$ in L^2. So the completion is in fact $H \oplus \mathbb{C}$ and the natural map of the completion into H is that which sends (f,z) into f.

There is an interesting application of the form representation theorem to a proof of a theorem of von Neumann. The idea is to consider a form which depends quadratically on the operator A.

<u>Proposition 2.2</u>. The operator $A : D(A) \to H$ is closed if and only if the form $S(f,g) = \langle Af, Ag\rangle$ on $D(A)$ is closed. The operator is closable if and only if the form is closable.

Proof: The mapping $f \to f \oplus Af$ from $\mathcal{D}(A)$ to $\mathcal{H} \oplus \mathcal{H}$ preserves inner products, and the injection of $\mathcal{D}(A)$ into $\mathcal{H}$ corresponds to projection onto the first summand.

Theorem 2.3 . Let A be a closed densely defined operator. Then A^*A is a positive self-adjoint operator.

Proof: The form $\langle Af,Ag\rangle$ on $\mathcal{D}(A)$ is positive, closed, and densely defined. By the form representation theorem the associated operator is positive and self-adjoint. But it is easy to check that this operator is just A^*A .

So far we have only considered forms of positive self-adjoint operators. We now turn to forms of arbitrary self-adjoint operators. Everything is as before - except that the convenient characterization of the forms is missing.

If $\mathcal{H}$ is a Hilbert space and A is a self-adjoint operator, its form domain $\mathcal{Q}(A) \subset \mathcal{H}$ is the form domain of the positive operator $|A|$. As before, the form domain $\mathcal{Q}(A)$ is larger than the operator domain $\mathcal{D}(A)$.

If $U : \mathcal{H} \to L^2(M,\mu)$ provides an isomorphism of A with multiplication by α , then $\mathcal{Q}(A)$ corresponds to $L^2(M,(1+|\alpha|)\mu)$. The form of A is given by $S(f,g) = \int (Uf)^*\alpha(Ug)d\mu$. This is clearly a Hermitian form. The associated operator is just $A : \mathcal{D}(A) \to \mathcal{H}$.

Let $\mathcal{Q} = \mathcal{Q}(A)$ and let $\mathcal{Q}^*$ be the conjugate dual space of the Hilbert space $\mathcal{Q}$. If g is in $\mathcal{H}$, then $f \mapsto \langle f,g\rangle$ is in $\mathcal{Q}^*$. Thus we may regard $\mathcal{H}$ as contained in $\mathcal{Q}^*$ and write $\langle f,v\rangle$ for the value of v in $\mathcal{Q}^*$ on f in $\mathcal{Q}$. In the spectral representation $\mathcal{Q}^*$ corresponds to $L^2(M,(1+|\alpha|)^{-1}d\mu)$.

If g is in $\mathcal{Q}$, we define Ag in $\mathcal{Q}^*$ by $\langle f,Ag\rangle = S(f,g)$ for f in $\mathcal{Q}$. This $A : \mathcal{Q} \to \mathcal{Q}^*$ is an extension of the original operator $A : \mathcal{D}(A) \to \mathcal{H}$.

The identifications $\mathcal{Q} \subset \mathcal{H} \subset \mathcal{Q}^*$ may be confusing. The point is that a Hilbert space is always isomorphic to its conjugate dual space. Our conventions are such that the isomorphism of $\mathcal{H}$ with $\mathcal{H}^*$ is the identity, while the isomorphism of $\mathcal{Q}$ with $\mathcal{Q}^*$ is $1 + |A|$.

One may define a notion of closed form in the general setting. Let $\mathcal{H}$ be a Hilbert space. Let $\mathcal{Q}$ be another Hilbert space such that $\mathcal{Q} \subset \mathcal{H}$, $\mathcal{Q}$ is dense in $\mathcal{H}$, and the injection is continuous. Let $\mathcal{Q}^*$ be the conjugate dual space of $\mathcal{Q}$ and identify $\mathcal{H} \subset \mathcal{Q}^*$. Let A be a bounded Hermitian operator from $\mathcal{Q}$ to $\mathcal{Q}^*$ and let S be the Hermitian form $S(f,g) = \langle f,Ag\rangle$. The Hermitian form S is said to be <u>closed</u> if $A - z : \mathcal{Q} \to \mathcal{Q}^*$ has range $\mathcal{Q}^*$ for some pair of complex conjugate z.

It is clear from Theorem 1.7 that the operator associated with a closed Hermitian form is self-adjoint. However with this definition it may be as difficult to check that the form is closed as to show directly that the operator is self-adjoint.

NOTES

The basic results on forms originate with Friedrichs. There is a very complete discussion in Chapter VI of the book of Kato (1966). The form representation theorem is there called the second representation theorem. (His first representation theorem is a slightly weaker result for a wider class of forms. The reason the result is weaker is related to the pathology of square roots of non-self-adjoint operators discussed in McIntosh (1972).) There is also a brief treatment in §7 of the notes of Nelson (1969).

The definition of closed Hermitian form in the general case is due to McIntosh (1970a) (1970b). He poses the question as to whether every closed Hermitian form is the form of its associated self-adjoint operator.

§3 THE ADDITION PROBLEM

The main subject of these lectures is the addition problem: When is the sum of two self-adjoint operators self-adjoint? In order to discuss this problem, we first need a definition of "sum". One possibility is the operator sum: $A + B$ has an obvious definition on $\mathcal{D}(A) \cap \mathcal{D}(B)$. However there is a more general notion.

Let A and B be self-adjoint operators acting in the Hilbert space $\mathcal{H}$. Consider their forms $\langle f,Ag \rangle$ on $\mathcal{Q}(A)$ and $\langle f,Bg \rangle$ on $\mathcal{Q}(B)$. Then $S(f,g) = \langle f,Ag \rangle + \langle f,Bg \rangle$ is a Hermitian form on $\mathcal{Q}(A) \cap \mathcal{Q}(B)$. Assume that $\mathcal{Q}(A) \cap \mathcal{Q}(B)$ is dense in $\mathcal{H}$ and let C be the operator associated with S . Then C is called the form sum of A and B .

Proposition 3.1 . Let C be the form sum of A and B . Then $\mathcal{D}(A) \cap \mathcal{D}(B) \subset \mathcal{D}(C)$.

Proof: Let g be in $\mathcal{D}(A) \cap \mathcal{D}(B)$. Then $S(f,g) = \langle f,Ag+Bg \rangle$, so g is in $\mathcal{D}(C)$ and $Cg = Ag + Bg$.

If C is the form sum of A and B we write $C = A + B$, even though in general it is possible that $\mathcal{D}(A) \cap \mathcal{D}(B) = \{0\}$.

The addition problem is important in quantum mechanics. We now summarize the principles of that subject.

A quantum mechanical system is specified by a family of self-adjoint operators acting in a Hilbert space $\mathcal{H}$; these correspond to the observables of the system. A state of the system is determined by a unit vector in $\mathcal{H}$.

If A is a self-adjoint operator, then by the spectral theorem there is a unitary transformation $U : \mathcal{H} \to L^2(M,\mu)$ such that A corresponds under U to multiplication by a real function α on M.

If f is a unit vector in $\mathcal{H}$, $\|f\| = 1$, then $\tilde{f} = Uf$ is an element of $L^2(M,\mu)$ with $\int |\tilde{f}|^2 d\mu = 1$. Thus $|\tilde{f}|^2 d\mu$ is a probability measure. The function α may then be regarded as a random variable on this probability space. Thus when the system is in the state determined by f, the probability that the observable corresponding to A has value in the set E of real numbers is $\int_{\alpha^{-1}(E)} |\tilde{f}|^2 d\mu$.

This probability may be expressed in a way which is independent of the representation by $\langle f, 1_E(A)f\rangle$, where $1_E(A)$ is the spectral projection of A corresponding to E.

Notice that α has an expectation if and only if f is in $\mathcal{Q}(A)$ and in that case the expectation is

$$m = \int \alpha |\tilde{f}|^2 d\mu = \langle f, Af\rangle .$$

It has a second moment if and only if f is in $\mathcal{D}(A)$, and the second moment is then

$$\int \alpha^2 |\tilde{f}|^2 d\mu = \|Af\|^2 .$$

The uncertainty (or standard deviation) of A is then $\Delta A = (\int (\alpha - m)^2 |\tilde{f}|^2 d\mu)^{\frac{1}{2}} = \|(A-m)f\|$.

Since the spectrum of A is the essential range of α, the probability that the value is in the spectrum is $\int |\tilde{f}|^2 d\mu = \langle f,f\rangle = 1$, whatever the state of the system.

The most important observable of a quantum mechanical system is the energy, since it determines the time development of the system. If H is the self-adjoint operator corresponding to the energy, by the spectral theorem it is isomorphic to multiplication by a real function. Thus for each real t, $\exp(-itH)$ is isomorphic to multiplication by a function of absolute value one, and hence is a unitary operator. If f determines the present state of the system, $\exp(-itH)f$ determines the state t units of time later.

If A and B are self-adjoint operators, they are isomorphic to multiplication operators, but not necessarily by the same unitary equivalence. Though in any state the corresponding observables may have probability distributions, there may be no determination of a joint distribution.

One important case of this is the kinetic energy H_o and the potential energy V. For each of these the isomorphism with a multiplication operator is explicitly known. They need not have a joint distribution, but the total energy H is supposed to be determined by adding expectations:

$$\langle f,Hf\rangle = \langle f,H_of\rangle + \langle f,Vf\rangle$$

for all f in $\mathcal{Q}(H_o)$ and $\mathcal{Q}(V)$. However it is necessary to find conditions on H_o and V so that this formula does indeed determine a self-adjoint operator H.

EXAMPLE

The Hilbert space which describes the states of a quantum mechanical particle moving in $\mathbb{R}^n$ may be taken as $\mathcal{H} = L^2(\mathbb{R}^n, dx)$. (Of course the case $n = 3$ is of primary importance, but it is illuminating to keep track of the dependence on n .)

The Fourier transformation gives an isomorphism of $\mathcal{H}$ with $L^2(\mathbb{R}^n, dk(2\pi)^{-n})$. The points k have the interpretation of momenta. In non-relativistic mechanics kinetic energy grows quadratically with momentum, like k^2 . The corresponding kinetic energy operator acting in $\mathcal{H}$ is $H_o = -\Delta$.

The points x have the interpretation of position. If V is a real measurable function on $\mathbb{R}^n$, $V(x)$ may be regarded as giving the potential energy at x . Multiplication by V is the potential energy operator acting in $\mathcal{H}$.

The natural question is then: For what V is $H = H_o + V$ self-adjoint? It is precisely for these V that the Schrödinger equation has a well-determined solution (given by $\exp(-itH)f$).

NOTES

Chernoff (1970), (1974) has investigated the general addition problem. He defines the sum of two self-adjoint operators by the Trotter product formula. He shows that when the sums exist the operation is commutative but not necessarily associative. The cancellation law also fails. He also shows that if an operator can be added to all self-adjoint operators, it must be bounded. And Goldstein (1972) has shown that the sum is discontinuous (with respect to strong resolvent convergence).

§4 POSITIVE FORM SUMS

The following theorem says that there is usually no difficulty adding positive self-adjoint operators.

Theorem 4.1 . Let $\mathcal{H}$ be a Hilbert space and H_o and U be positive self-adjoint operators acting in $\mathcal{H}$. Assume that $\mathcal{Q}(H_o) \cap \mathcal{Q}(U)$ is dense in $\mathcal{H}$. Then the form sum $H = H_o + U$ is a positive self-adjoint operator with $\mathcal{Q}(H) = \mathcal{Q}(H_o) \cap \mathcal{Q}(U)$.

Proof: Let $\mathcal{Q} = \mathcal{Q}(H_o) \cap \mathcal{Q}(U)$. The form $S(f,g) = \langle f,H_o g\rangle + \langle f,Ug\rangle$ is a closed positive form on $\mathcal{Q}$. By the form representation theorem it is the form of a positive self-adjoint operator H with $\mathcal{Q}(H) = \mathcal{Q}$.

This theorem has an obvious generalization to operators which are bounded below. (Just add a constant to each operator.)

EXAMPLE

Let $\mathcal{H} = L^2(\mathbb{R}^n,dx)$. Let $H_o = -\Delta$ and let U be a real measurable function on $\mathbb{R}^n$ which is bounded below. Assume that U is in L^1 locally on the complement of a closed set of zero measure. Then $H = H_o + U$ is self-adjoint and bounded below and $\mathcal{Q}(H) = \mathcal{Q}(H_o) \cap \mathcal{Q}(U)$.

To see this it is sufficient to show that $\mathcal{Q}(H_o) \cap \mathcal{Q}(U)$ is dense in $\mathcal{H}$. We have assumed that U is locally integrable on the complement of a closed set M of measure zero. Let $\mathcal{F}$ be the space of C^∞ functions with compact support in the complement of M . Clearly $\mathcal{F} \subset \mathcal{Q}(H_o) \cap \mathcal{Q}(U)$ and $\mathcal{F}$ is dense in $\mathcal{H}$.

NOTES

Sums of forms are discussed in Chapter VI of the book by Kato (1966). (There is often confusion about the relation of the form sum to the Friedrichs extension. Kato shows on page 329 that the form sum need not be the Friedrichs extension of the operator sum.)

§5 SMALL FORM PERTURBATIONS

Another technique for defining sums of self-adjoint operators is perturbation theory. An operator of either sign can be added on, provided that it is relatively small. In this case self-adjointness of the sum is proved by means of a convergent geometric series expansion.

In order to obtain convergence, it is necessary to have a notion of what is meant by relatively small. For this we need some basic facts about norms.

Let E be a complex vector space with inner product $\langle f,g \rangle$ and norm $\|f\| = \langle f,f \rangle^{\frac{1}{2}}$. Let $S(f,g)$ be another sesquilinear form defined on E. Define the sesquilinear form norm by $\|S\| = \sup\{|S(f,g)| : \|f\| \leq 1, \|g\| \leq 1\}$ and the quadratic form norm by $w(S) = \sup\{|S(f,f)| : \|f\| \leq 1\}$. The sesquilinear form is said to be bounded if its sesquilinear form norm is finite.

Proposition 5.1. It is always true that $w(S) \leq \|S\| \leq 2w(S)$. If S is Hermitian, then $w(S) = \|S\|$.

Proof: It is clear that $w(S) \leq \|S\|$.

On the other hand, we see from the polarization identity that

$$4|S(f,g)| \leq w(S)(\|f+g\|^2+\|f-g\|^2+\|f+ig\|^2+\|f-ig\|^2) = 4w(S)(\|f\|^2+\|g\|^2) .$$

This implies that $\|S\| \leq 2w(S)$.

If S is Hermitian, then $S(h,h)$ is always real. Hence $4\mathrm{Re}S(f,g) = S(f+g,f+g) - S(f-g,f-g)$. It follows that

$$4|\mathrm{Re}S(f,g)| \leq w(S)(\|f+g\|^2+\|f-g\|^2) = 2w(S)(\|f\|^2+\|g\|^2) .$$

This implies that $\|S\| \leqslant w(S)$.

Let E and F be inner product spaces. The <u>norm</u> of a linear (or conjugate linear) transformation $A : E \to F$ is $\|A\| = \sup\{\|Af\| : \|f\| \leqslant 1\}$. The transformation is said to be <u>bounded</u> if its norm is finite. It is easy to see that A is bounded if and only if it is continuous.

Recall that if E is an inner product space, its conjugate dual space E^* consists of the bounded conjugate linear functionals on E . The value of v in E^* on f in E is denoted $\langle f,v\rangle$. There is a natural bijective norm preserving correspondence between bounded sesqui-linear forms S on E and bounded operators $A : E \to E^*$ given by $S(f,g) = \langle f,Ag\rangle$.

<u>Theorem 5.2</u> . Let H_o be a self-adjoint operator acting in the Hilbert space $\mathcal{H}$. Let W be a Hermitian form on $\mathcal{Q}(H_o)$ such that there exist positive constants a and b with $a < 1$ such that $\pm\langle f,Wf\rangle \leqslant a\langle f,(|H_o|+b)f\rangle$ for all f in $\mathcal{Q}(H_o)$. Then the form sum $H = H_o + V$ is self-adjoint.

<u>Proof</u>: Assume $b > 0$ and give $\mathcal{Q}(H_o)$ the norm $\|f\|_{\mathcal{Q}} = \|(|H_o|+b)^{\frac{1}{2}}f\|$.

We have $\mathcal{Q} \subset \mathcal{H} \subset \mathcal{Q}^*$ and $H = H_o + W : \mathcal{Q} \to \mathcal{Q}^*$. Define the Hermitian operator H acting in $\mathcal{H}$ by restricting to $\mathcal{D}(H) = \{f \text{ in } \mathcal{Q} : Hf \text{ is in } \mathcal{H}\}$. To show that $H : \mathcal{D}(H) \to \mathcal{H}$ is self-adjoint, it is sufficient to show that range $(H-z) = \mathcal{H}$ for z imaginary and sufficiently large. For this it is sufficient to show that $H : \mathcal{Q} \to \mathcal{Q}^*$ satisfies range $(H-z) = \mathcal{Q}^*$.

This would be established if we could show the existence of $(H-z)^{-1} = [1+(H_o-z)^{-1}W]^{-1}(H_o-z)^{-1} : \mathcal{Q}^* \to \mathcal{Q}$ by a series expansion. For this it is enough to show that $(H_o-z)^{-1}W : \mathcal{Q} \to \mathcal{Q}$ has norm strictly less than one.

But by assumption $\pm\langle f,Wf\rangle \leq a\|f\|_{\mathcal{Q}}^2$. It follows that $|\langle f,Wg\rangle| \leq a\|f\|_{\mathcal{Q}}\|g\|_{\mathcal{Q}}$, that is, $W : \mathcal{Q} \to \mathcal{Q}^*$ is an operator with norm bounded by a . On the other hand $|H_o| + b : \mathcal{Q} \to \mathcal{Q}^*$ is an isomorphism of Hilbert spaces, so $(|H_o|+b)^{-1}W : \mathcal{Q} \to \mathcal{Q}$ has norm bounded by a . Thus the norm of
$(H_o-z)^{-1}W = (H_o-z)^{-1}(|H_o|+b)(|H_o|+b)^{-1}W : \mathcal{Q} \to \mathcal{Q}$ is bounded by $(1+b/|z|)a < 1$, for $|z|$ sufficiently large.

EXAMPLES

Let $\mathcal{H} = L^2(\mathbb{R}^n,dx)$. Let $H_o = -\Delta$ and let $V = U + W$, where U and W are real measurable functions on $\mathbb{R}^n$. Assume U is bounded below and that $\mathcal{Q}(H_o) \cap \mathcal{Q}(U)$ is dense in $\mathcal{H}$. Assume that W is in $L^p + L^\infty$ for some $p \geq \frac{n}{2}$ (and where $p \geq 1$ if $n = 1$, $p > 1$ if $n = 2$). Then $H = -\Delta + V$ is self-adjoint and bounded below and $\mathcal{Q}(H) = \mathcal{Q}(H_o) \cap \mathcal{Q}(V)$.

All that is needed to prove this is an estimate $\pm W \leq a(H_o+b)$ for some $a < 1$. For since U is bounded below it follows that $\pm W \leq a(H_o+U+c)$ and so W is a small form perturbation of $H_o + U$.

Write $W = W_\infty + W_p$, where W_∞ is bounded and W_p is in L^p . The L^p norm of W_p may be chosen arbitrarily small. But then by Hölder's inequality

$$|\langle f,W_pf\rangle| \leq \|W_p\|_p \, \|f^2\|_q = \|W\|_p\|f\|_{2q}^2 \quad \text{where } \frac{1}{p} + \frac{1}{q} = 1 \, .$$

Notice that $\frac{1}{2q} \geq \frac{1}{2} - \frac{1}{n}$. Thus the bound on $\|f\|_{2q}^2$ in terms of $\langle f,(-\Delta+b)f\rangle$ follows from Sobolev inequalities.

We see in the next example that for $H_o + W$ to be a self-adjoint operator it is not at all necessary for the form W to be the form of a self-adjoint operator.

Let $\mathcal{H} = L^2(\mathbb{R},dx)$ and $A = -\frac{d^2}{dx^2}$. Recall that if f is in $\mathcal{Q}(A)$, then its Fourier transform $\hat{f}$ is in L^1 . Let $\hat{\mu}$ be a function in L^∞ with $\hat{\mu}(-k) = \hat{\mu}(k)^*$. Define $W : \mathcal{Q} \rightarrow \mathcal{Q}^*$ by $\langle f,Wg\rangle = \iint \hat{f}(k)^*\hat{\mu}(k-r)\hat{g}(r)dk/2\pi\, dr/2\pi$. Then W is a Hermitian form which satisfies the hypotheses of the theorem. Hence $-\frac{d^2}{dx^2} + W$ is a self-adjoint operator acting in $\mathcal{H}$.

If $\hat{\mu}$ is the Fourier transform of a real function μ in L^1 , then W is multiplication by μ , and hence is a self-adjoint operator. However there are cases where W is not a multiplication operator.

For instance, let b be a real number and p a point on the line and set $\hat{\mu}(k) = b \exp(ipk)$. Then $\langle f,Wg\rangle = bf(p)^*g(p)$. Thus $W : \mathcal{Q} \rightarrow \mathcal{Q}^*$ is given by $Wg = bg\delta_p$, where δ_p is the Dirac measure at p defined by $\langle f,\delta_p\rangle = f(p)^*$.

In this example, the functions g in the domain of the self-adjoint restriction $A + W = -\frac{d^2}{dx^2} + b\delta_p$ have a slope discontinuity at p equal to $bf(p)$. Thus, for such g , $(A+W)g$ is in L^2 even though it is false in general that Ag or Wg is in L^2 .

The resolvent of $A + W$ can be computed explicitly. For convenience, set $p = 0$. Then

$$(A+W+c)^{-1}g(x) = (A+c)^{-1}g(x)$$

$$- (b/(2c^{\frac{1}{2}}+b))\exp(-c^{\frac{1}{2}}|x|) \int \frac{1}{2c^{\frac{1}{2}}} \exp(-c^{\frac{1}{2}}|y|)g(y)dy .$$

If $b < 0$ then $(A+W+c)^{-1}$ has a pole at $c = \frac{b^2}{4}$. Thus the self-adjoint operator has an eigenvalue $-\frac{b^2}{4}$. The corresponding eigenfunction is $\exp(\frac{1}{2}b|x|)$.

Another example is $\hat{\mu}(k) = \pi i \operatorname{sign}(k)$. To analyze this, note that the inverse Fourier transform of $\pi i \exp(-\varepsilon|k|)\operatorname{sign}(k)$ is $\frac{x}{x^2+\varepsilon^2}$, for $\varepsilon > 0$. Hence

$$\langle f,Wg\rangle = \lim_{\varepsilon \downarrow 0} \int f(x)^* \frac{x}{x^2+\varepsilon^2} g(x)dx = \text{p.v.} \int f(x)^* \frac{1}{x} g(x)dx .$$

This principal value integral is <u>not</u> the form associated with multiplication by $\frac{1}{x}$.

NOTES

Theorem 5.2 is related to what Kato (1966) calls the pseudo-Friedrichs extension. However it is really not so much an extension theorem as a perturbation theorem.

The theorem may not be strengthened to allow $a = 1$. For if $A = -\frac{d^2}{dx^2}$ and $H_0 = A + \delta$, $W = -A$, then $\pm W \leq H_0$ but $H_0 + W = \delta$ is not a closable form.

Sobolev inequalities are treated in Stein (1970).

The class of Schrödinger operators for which the form sum is self-adjoint and bounded below has most recently been studied in detail by Schechter (1972).

§6 SMALL OPERATOR PERTURBATIONS

The more traditional approach to perturbation theory is in terms of operator sums rather than form sums. The hypotheses of the theorems are stronger (second order rather than first order in the operators), but so are the conclusions.

We begin by reviewing the relation between first order and second order estimates. First we need a preliminary remark. Then we can discuss the order relation between self-adjoint operators.

If $\mathcal{H}$ is a Hilbert space and $C : \mathcal{H} \to \mathcal{H}$ is a bounded operator, its norm is the norm of the form $\langle f, Cg \rangle$. The form of the adjoint $C^* : \mathcal{H} \to \mathcal{H}$ is $\langle f, C^* g \rangle = \langle Cf, g \rangle = \langle g, Cf \rangle^*$. Thus C^* is also bounded and in fact $\|C^*\| = \|C\|$.

Let A and B be self-adjoint operators. Then $B \leqslant A$ means that $\mathcal{Q}(A) \subset \mathcal{Q}(B)$ and $\langle f, Bf \rangle \leqslant \langle f, Af \rangle$ for all f in $\mathcal{Q}(A)$.

Notice that $B^2 \leqslant A^2$ means that $\mathcal{D}(B) \supset \mathcal{D}(A)$ and $\|Bf\| \leqslant \|Af\|$.

Proposition 6.1. Let A and B be self-adjoint operators which are bounded below and let c be a real number which is strictly less than the lower bounds. Then $B \leqslant A$ if and only if $(A-c)^{-1} \leqslant (B-c)^{-1}$.

Proof: By adding a constant we may assume $c = 0$. Then $B \leqslant A$ says that $\|B^{\frac{1}{2}}f\|^2 \leqslant \|A^{\frac{1}{2}}f\|^2$, that is, $\|B^{\frac{1}{2}}A^{-\frac{1}{2}}\| \leqslant 1$. But $(B^{\frac{1}{2}}A^{-\frac{1}{2}})^* \supset A^{-\frac{1}{2}}B^{\frac{1}{2}}$, so $\|A^{-\frac{1}{2}}B^{\frac{1}{2}}\| \leqslant 1$. That is, $\|A^{-\frac{1}{2}}g\| \leqslant \|B^{-\frac{1}{2}}g\|$, $A^{-1} \leqslant B^{-1}$.

The next proposition is the main result about this order relation: second order estimates imply first order estimates.

Proposition 6.2. Let A and B be positive self-adjoint operators with $0 \leqslant B^2 \leqslant A^2$. Then $0 \leqslant B \leqslant A$.

Proof: First consider the special case when there exists a c with $0 < c \leq B^2 \leq A^2$. Then $(A^2+t^2)^{-1} \leq (B^2+t^2)^{-1}$, so using $A^{-1} = \frac{2}{\pi}\int_0^\infty (A^2+t^2)^{-1}dt$ and the corresponding representation for B^{-1} we see that $A^{-1} \leq B^{-1}$, and so $B \leq A$.

In the general case we have $0 \leq B \leq (B^2+\varepsilon)^{\frac{1}{2}} \leq (A^2+\varepsilon)^{\frac{1}{2}}$ for every $\varepsilon > 0$. Hence $B \leq A$, by the monotone convergence theorem.

Theorem 6.3. Let H_o be a self-adjoint operator acting in the Hilbert space $\mathcal{H}$. Let W be a self-adjoint operator acting in $\mathcal{H}$ with $\mathcal{D}(W) \supset \mathcal{D}(H_o)$. Assume that there exist constants a and b with $a^2 < 1$ such that $W^2 \leq a^2(H_o^2+b^2)$. Then the operator sum $H = H_o + W$ is a self-adjoint operator with $\mathcal{D}(H) = \mathcal{D}(H_o)$.

Proof: Assume $b \neq 0$ and set $z = \pm ib$. Then

$$\|W(H_o-z)^{-1}f\|^2 \leq a^2\langle(H_o-z)^{-1}f,(H_o^2+b^2)(H_o-z)^{-1}f\rangle = a^2\|f\|^2 .$$

Hence $W(H_o-z)^{-1} : \mathcal{H} \to \mathcal{H}$ has norm bounded by $a^2 < 1$ and $(H-z)^{-1} = (H_o-z)^{-1}[1+W(H_o-z)^{-1}]^{-1}$ exists. Thus $H = H_o + W : \mathcal{D}(H_o) \to \mathcal{H}$ is self-adjoint.

Notice that $W^2 \leq a^2(H_o^2+b^2)$ implies $W^2 \leq a^2(|H_o|+b)^2$ which in turn implies that $\pm W \leq a(|H_o|+b)$. Thus the hypotheses of Theorem 6.3 are stronger than those of Theorem 5.2. But the conclusion is also, since we can identify $\mathcal{D}(H)$.

EXAMPLES

Let $\mathcal{H} = L^2(\mathbb{R}^n,dx)$ and $H_o = -\Delta$. Let W be a real measurable function on $\mathbb{R}^n$. Assume that W is in L^p for some $p \geq \frac{n}{2}$ (and

$p \geqslant 2$ if $n = 1,2$, or 3 , $p > 2$ if $n = 4$). Then $H = H_o + W$ is self-adjoint with $\mathcal{D}(H) = \mathcal{D}(H_o)$.

Write $W = W_\infty + W_p$, where W_∞ is bounded and W_p is in L^p . The L^p norm of W_p may be chosen arbitrarily small. By Hölder's inequality $\|Wf\|_2 = \|W_p\|_p \|f\|_q$, where $\frac{1}{p} + \frac{1}{q} = \frac{1}{2}$. Notice that $\frac{1}{q} \geqslant \frac{1}{2} - \frac{2}{n}$. The bound on $\|f\|_q$ in terms of $\|(-\Delta+b)f\|$ follows from Sobolev inequalities.

Next we turn to a class of results where the domain of the self-adjoint operator is not determined quite so explicitly.

Proposition 6.4 . Let A be a Hermitian operator. Then A is closable and $\bar{A}$ is also Hermitian.

Proof: Since $A \subset A^*$ and A^* is closed, A is closable and $\bar{A} \subset A^*$.

Proposition 6.5 . Let A be Hermitian and A_1 be self-adjoint with $A \subset A_1$. Then $\bar{A} \subset A_1 \subset A^*$.

Proof: Since $A_1 = A_1^*$, A_1 is closed. Thus $\bar{A} \subset A_1$. Since $A \subset A_1$, $A_1^* \subset A^*$, that is, $A_1 \subset A^*$.

A Hermitian operator A is said to be essentially self-adjoint if $\bar{A}$ is self-adjoint.

Since $\bar{A} = A^{**}$, A Hermitian operator A is essentially self-adjoint $(A^* \subset \bar{A})$ if and only if A^* is Hermitian $(A^* \subset A^{**})$.

Proposition 6.6 . If A is essentially self-adjoint, then it has a unique self-adjoint extension.

Proof: If A_1 is self-adjoint, then $\bar{A} \subset A_1 \subset A^* = \bar{A}^* = \bar{A}$.

EXAMPLES

(1) Let $\mathcal{H} = L^2(\mathbb{R}^n, dx)$ and let A be the Laplace operator with domain $\mathcal{D}(A) = C^\infty_{com}(\mathbb{R}^n)$, the space of C^∞ functions with compact support. Then A is essentially self-adjoint, and its closure is the self-adjoint Laplace operator Δ.

The following proof of this result uses an important technique, invariance under a unitary group. Let f be in C^∞_{com}. Since C^∞_{com} is translation invariant, the translates f_a of f are also in C^∞_{com}. Let g be orthogonal to C^∞_{com} in the inner product of $\mathcal{D}(\Delta)$. Thus $\langle \Delta g, \Delta f_a\rangle + \langle g, f_a\rangle = 0$. It follows that

$$\langle \hat{g}, (1+k^4)\hat{f}_a\rangle = \langle \hat{g}, (1+k^4)\exp(ika)\hat{f}\rangle = 0 .$$

Since the Fourier transform of an L^1 function determines the function, $\hat{g}*(1+k^4)\hat{f} = 0$. Since f may be chosen so that $\hat{f}$ is never zero, it follows that $\hat{g} = 0$, $g = 0$.

(2) Let $\mathcal{H} = L^2(\mathbb{R}^n, dx)$ and let $A = -\Delta$ be the Laplace operator. Let $\mathcal{E} = C^\infty_{com}(\mathbb{R}^n - 0)$, the space of C^∞ functions with compact support not intersecting the origin. We ask when $\mathcal{E}$ is dense in $\mathcal{Q}(A)$ and when it is dense in $\mathcal{D}(A)$. (When the latter holds, A restricted to $\mathcal{E}$ is essentially self-adjoint.) In order to discuss this question it will be convenient to introduce the language of tempered distributions.

Let $\mathcal{S}$ be the space of rapidly decreasing C^∞ functions on $\mathbb{R}^n$. (<u>Rapidly decreasing</u> means that each derivative goes to zero at infinity faster than every polynomial in $\frac{1}{|x|}$.) A sequence g_n of elements of $\mathcal{S}$ is said to converge to g in $\mathcal{S}$ if for every p and q, $x^q D^p g_n \to x^q D^p g$ uniformly.

A tempered distribution is a continuous conjugate linear functional on S . If v is a tempered distribution and g is in S , the value of v on g will be written $\langle g,v\rangle$.

If u is a function on $\mathbb{R}^n$ such that, for some m , $(1+x^2)^{-m}u$ is integrable, then u defines a tempered distribution by $\langle g,u\rangle = \int g^* u dx$. It is a fact that this tempered distribution determines the function (almost everywhere).

The nice features of tempered distributions stem from the fact that the Fourier transform maps S onto S . Thus if v is a tempered distribution, the Fourier transform $\hat{v}$ of v may be defined by $\langle \hat{g},\hat{v}\rangle = \langle g,v\rangle$, where g is an arbitrary element of S . Then $\hat{v}$ is also a tempered distribution.

For example, the Fourier transform of the Dirac measure δ defined by $\langle g,\delta\rangle = g(0)^*$ is given by $\langle \hat{g},\hat{\delta}\rangle = g(0)^* = \langle \hat{g},1\rangle$. Thus $\hat{\delta}$ corresponds to the function 1 .

Differentiation sends S into S . Thus if v is a tempered distribution, its derivative $D^p v$ may be defined by $\langle g,D^p v\rangle = (-1)^{|p|}\langle D^p g,v\rangle$. This may be expressed in terms of Fourier transforms as $\langle g,D^p v\rangle = \langle (-ik)^p \hat{g},\hat{v}\rangle$. For example the Fourier transform of $D^p\delta$ corresponds to the polynomial $(ik)^p$.

If $H = L^2(\mathbb{R}^n,dx)$ and $A = -\Delta$, then Q consists of the functions f in H such that $\int |f|^2(1+k^2)dk < \infty$. It is clear from this that S is dense in Q . Hence the elements of Q^* may be identified with certain tempered distributions v , in fact those whose Fourier transforms correspond to functions $\hat{v}$ such that $\int |\hat{v}|^2(1+k^2)^{-1}dk < \infty$. The value of v on f is then $\langle f,v\rangle = \langle \hat{f},\hat{v}\rangle = \int \hat{f}^*\hat{v}dk(2\pi)^{-n}$. The Fourier transforms of the elements of $\mathcal{Q}^*$ are thus functions, even if the elements of $\mathcal{Q}^*$ are only tempered distributions.

Similarly, $\mathcal{D}$ consists of the functions such that $\int |\hat{f}|^2(1+k^4)dk < \infty$. Its conjugate dual space $\mathcal{D}^*$ may be identified with those tempered distributions which have Fourier transforms $\hat{v}$ satisfying $\int |\hat{v}|^2(1+k^4)^{-1}dk < \infty$.

In order to see when E is dense in $\mathcal{Q}$, consider v in $\mathcal{Q}^*$ with $\langle f,v\rangle = 0$ for all f in E. It is sufficient to show that $v = 0$. But since v is a distribution with support at one point, it is a linear combination of derivatives of a δ measure. That is, $\hat{v}$ is a polynomial. But v is also in $\mathcal{Q}^*$, so $\int |\hat{v}(k)|^2(1+k^2)^{-1}dk < \infty$. If $n \geq 2$ this implies $v = 0$

To see when E is dense in $\mathcal{D}$, we use the same argument. If v in $\mathcal{D}^*$ is zero on E, then $\hat{v}$ is a polynomial. But since $\int |\hat{v}|^2(1+k^4)^{-1}dk < \infty$, $\hat{v}$ must be zero if $n \geq 4$.

If A is an operator, its numerical range consists of the set of complex numbers $\langle g,Ag\rangle$ such that g is in $\mathcal{D}(A)$ with $\|g\| = 1$.

If A is a Hermitian operator, its numerical range is a subset of the real numbers.

<u>Proposition 6.7</u>. Let A be an operator and let z be a complex number which is at a strictly positive distance d from the numerical range of A. Then $A - z$ is injective and $\|(A-z)^{-1}f\| \leq (1/d)\|f\|$ for all f in range $(A-z)$.

<u>Proof</u>: Let g be in $\mathcal{D}(A)$ with $\|g\| = 1$. Then $\|(A-z)g\| \geq |\langle g,(A-z)g\rangle| = |\langle g,Ag\rangle - z| \geq d$. Thus for all h in $\mathcal{D}(A)$, $\|(A-z)h\| \geq d\|h\|$. Let $f = (A-z)h$. This gives the estimate

Proposition 6.8 . Let A be a Hermitian operator and let z be a complex number which is at a strictly positive distance from the numerical range of A . Assume that range $A - z$ and range $A - z^*$ are dense in $\mathcal{H}$. Then A is essentially self-adjoint.

Proof: Let $\bar{A}$ be the closure of A . Then z is at a strictly positive distance from the numerical range of $\bar{A}$. Hence by Proposition 6.7 $(\bar{A}-z)^{-1}$ and $(\bar{A}-z^*)^{-1}$ are continuous. It follows from the Corollary to Proposition 1.6 that range $\bar{A} - z$ and range $\bar{A} - z^*$ are closed. Since they are dense by assumption, they are both equal to $\mathcal{H}$. Theorem 1.7 implies that $\bar{A}$ is self-adjoint.

Theorem 6.9 . Let H_o be a self-adjoint operator acting in the Hilbert space $\mathcal{H}$. Let W be a self-adjoint operator acting in $\mathcal{H}$ with $\mathcal{D}(W) \supset \mathcal{D}(H_o)$. Assume that there exists a constant b such that $W^2 \leqslant H_o^2 + b^2$. Then the operator $H = H_o + W$ with domain $\mathcal{D}(H_o)$ is essentially self-adjoint.

Proof: Assume $b \neq 0$ and set $z = \pm ib$. Then $T = W(H_o-z)^{-1}$ has norm bounded by one. Write $H - z = [1+T](H_o-z)$. It is sufficient to show that range $(1+T)$ is dense in $\mathcal{H}$.

If f is orthogonal to range $(1+T)$, then in particular it is orthogonal to $(1+T)f$. Since $(1+T)f - f = Tf$, we have

$$\|(1+T)f\|^2 + \|f\|^2 = \|Tf\|^2 \leqslant \|f\|^2 .$$

Hence $(1+T)f = 0$.

Next we show that $(1+T)f = 0$ implies $f = 0$. Let $g = (H_o-z)^{-1}f$. Then $(H-z)g = (1+T)f = 0$. Since H is Hermitian and z is not real, $g = 0$. Hence $f = 0$.

Thus the only element orthogonal to range (1+T) is $f = 0$. So range (1+T) is dense.

NOTES

There is an alternative proof of Proposition 6.2 in Kato's book (1966) (on page 292).

Theorem 6.3 is the classic perturbation theorem of Rellich.

Theorem 6.9 is taken from Chapter V of Kato's book (1966). Wüst (1971) showed that the hypothesis $\|Wf\|^2 \leqslant \|H_o f\|^2 + b^2\|f\|^2$ may be replaced by $\|Wf\| \leqslant \|H_o f\| + b\|f\|$. (This has been generalized to Banach spaces by Chernoff (1972a) and Okazawa (1971).) Wüst (1972) also has other results on borderline cases.

There is a discussion of tempered distributions in the book by Stein and Weiss (1971). The characterization of distributions with support at a point may be found in the treatise of Schwartz (1966).

§7 STRONG CONVERGENCE AND APPROXIMATION

In previous sections we have seen that the sum of two self-adjoint operators is defined under rather general conditions. For instance, if H_o and U are positive self-adjoint operators, then $H = H_o + U$ is self-adjoint in circumstances where U is far from being a small perturbation of H_o in any sense. The question arises whether such a sum may be approximated by small perturbations. Thus there would be a possibility of calculating approximations to the resolvent and other functions of H.

The purpose of this section is to show that this can be done. We construct a sequence U_k of bounded operators which approximate U. The resolvent of $H_k = H_o + U_k$ may be computed by an uniformly convergent series expansion in powers of $U_k(H_o-z)^{-1}$. This resolvent is then supposed to converge strongly (that is, pointwise) to the resolvent of H.

The basic problem is thus to develop techniques to prove strong convergence. We begin with a review of the various kinds of convergence and the basic facts about them. This will lead to a series of results about when weak convergence and certain order relationships imply strong convergence.

Consider a Hilbert space $\mathcal{H}$. There are two important notions of convergence for elements of $\mathcal{H}$. We say $f_n \to f$ <u>strongly</u> if the norm $\|f_n-f\| \to 0$. We say $f_n \to f$ <u>weakly</u> if $\langle g,f_n\rangle \to \langle g,f\rangle$ for all g in $\mathcal{H}$. Strong convergence obviously implies weak convergence.

Proposition 7.1 . If $f_n \to f$ weakly, and if $\|f\| \geq \limsup \|f_n\|$, then $f_n \to f$ strongly. (Weak convergence and no loss of norm implies strong convergence.)

Proof: Then $\|f_n - f\|^2 = \|f_n\|^2 - \langle f_n, f\rangle - \langle f, f_n\rangle + \|f\|^2$ has lim sup less than $\|f\|^2 - 2\|f\|^2 + \|f\|^2 = 0$.

There are at least three important types of convergence for bounded operators $A : \mathcal{H} \to \mathcal{H}$. We say $A_n \to A$ uniformly (on bounded sets) if the norm $\|A_n - A\| \to 0$. We say $A_n \to A$ strongly if $A_n f \to Af$ strongly for each f in $\mathcal{H}$. We say $A_n \to A$ weakly if $A_n f \to Af$ weakly for each f in $\mathcal{H}$, that is, the numbers $\langle g, A_n f\rangle$ converge to $\langle g, Af\rangle$ for each f and g in $\mathcal{H}$. Uniform convergence implies strong convergence which in turn implies weak convergence.

The following four propositions deal with convergence of functions of bounded operators. The first two are elementary.

Proposition 7.2 . If $A_n \to A$ strongly, $B_n \to B$ strongly, and there exists a constant M with $\|A_n\| \leq M$ for all n , then $A_n B_n \to AB$ strongly. (Multiplication is strongly continuous on bounded sets.)

Proposition 7.3 . If $A_n \to A$ weakly, then $A_n^* \to A^*$ weakly. (The adjoint is weakly continuous.)

A bounded normal operator is a bounded operator such that $AA^* = A^*A$. (In other words, A is normal if $\|A^*g\| = \|Ag\|$ for all g in $\mathcal{H}$.

Proposition 7.4 . Let A_n and A be bounded normal operators. If $A_n \to A$ strongly, then $A_n^* \to A^*$ strongly.

Proof: If $A_n \to A$ strongly, then $A_n \to A$ weakly, so $A_n^* \to A^*$ weakly. That is, $A_n^* g \to A^* g$ weakly for all g in H. But since A_n and A are normal, $\|A_n^* g\| = \|A_n g\| \to \|Ag\| = \|A^* g\|$. Thus there is no loss of norm, so $A_n^* g \to A^* g$ strongly. Since this holds for all g in H, $A_n^* \to A^*$ strongly as operators.

The spectral theorem for bounded normal operators states that any such operator is isomorphic to multiplication by a bounded complex function on some L^2 space. A Borel measurable subset S of the complex plane will be said to be of spectral measure zero if the set on which this function takes values in S is of measure zero.

Proposition 7.5. Let R_n be a sequence of bounded normal operators and let R be a bounded normal operator such that $R_n \to R$ strongly. Let ϕ be a bounded Borel measurable function on the complex plane which is continuous except on a closed set of R spectral measure zero. Then $\phi(R_n) \to \phi(R)$ strongly.

Proof: By the principle of uniform boundedness the $\|R_n\|$ are bounded. Thus it is sufficient to consider ϕ on a compact disc D in the complex plane.

First we show that the result is true for any ϕ which is continuous on D. The set of ϕ for which the result is true is a uniformly closed algebra, by Proposition 7.2. It contains z by assumption. Further, it is closed under complex conjugation, by Proposition 7.4. Hence by the Weierstrass approximation theorem it contains all continuous functions.

Next we show that the result extends to certain discontinuous functions ϕ. Let γ be a continuous function which is zero on the discontinuities of ϕ . Since both γ and $\phi\gamma$ are continuous, $\phi(R_n)\gamma(R) = \phi(R_n)[\gamma(R)-\gamma(R_n)] + \phi(R_n)\gamma(R_n)$ converges strongly to $\phi(R)\gamma(R)$. In other words, $\phi(R_n)g \to \phi(R)g$ for all vectors g of the form $g = \gamma(R)f$. Thus it is sufficient to show that such vectors are dense in $\mathcal{H}$.

To see this, choose γ_n to be a continuous function which is 0 on the discontinuities of ϕ and 1 at a distance greater than $\frac{1}{n}$ from them. Since the discontinuities are of spectral measure zero, the dominated convergence theorem implies that $\gamma(R_n)f \to f$ for all f in $\mathcal{H}$.

The next two propositions show how one can discuss convergence of functions of self-adjoint operators.

Let $\mathcal{H}$ be a Hilbert space. Let A_n be a sequence of self-adjoint operators acting in $\mathcal{H}$. We say that the A_n converge to A in the sense of <u>strong resolvent convergence</u> if for some z (bounded away from the spectra of the A_n and A) $(A_n-z)^{-1} \to (A-z)^{-1}$ strongly.

<u>Proposition 7.6</u> . Let A_n be a sequence of self-adjoint operators. Let A be another self-adjoint operator. Assume that there is a subspace $\mathcal{E} \subset \mathcal{D}(A_n)$ such that $A_n f \to Af$ for all f in $\mathcal{E}$. If $\mathcal{E}$ is dense in $\mathcal{D}(A)$, then $A_n \to A$ in the sense of strong resolvent convergence.

<u>Proof</u>: Let f be in $\mathcal{E}$ and set $h = (A-z)f$. Then
$(A-z)^{-1}h - (A_n-z)^{-1}h = (A_n-z)^{-1}(A_n-A)(A-z)^{-1}h = (A_n-z)^{-1}(A_n-A)f \to 0$. But since $\mathcal{E}$ is dense in $\mathcal{D}(A)$, $(A-z)\mathcal{E}$ is dense in $\mathcal{H}$. Since we have convergence on a dense set and a uniform bound, we have convergence on all of $\mathcal{H}$.

If ϕ is a bounded Borel measurable function on the real line, and A is a self-adjoint operator, then $\phi(A)$ is a bounded normal operator. (Of course how ϕ is defined off the spectrum of A doesn't matter.)

A Borel measurable subset E of the line will be said to be of <u>spectral measure zero</u> (with respect to the self-adjoint operator A) if $1_E(A) = 0$.

<u>Proposition 7.7</u> . Assume that the sequence A_n of self-adjoint operators converges to the self-adjoint operator A in the sense of strong resolvent convergence. Let ϕ be a bounded Borel function on the real line which is continuous except on a closed set of A spectral measure zero. Then $\phi(A_n) \to \phi(A)$ strongly.

<u>Proof</u>: Let $\psi(t) = \phi(z+1/t)$ and set $R = (A-z)^{-1}$, so that $\psi(R) = \phi(A)$. Then ψ is continuous except on a set of R spectral measure zero (which may include zero). Thus if we set $R_n = (A_n-z)^{-1}$, we have $\phi(A_n) = \psi(R_n) \to \psi(R) = \phi(A)$.

The next three results are concerned with when weak convergence (that is, convergence of forms) implies strong convergence. The conditions are stated in terms of the order relation between self-adjoint operators.

<u>Proposition 7.8</u> . Let R_n and R be bounded self-adjoint operators with $R_n \leq R$. If $R_n \to R$ weakly, then $R_n \to R$ strongly.

<u>Proof</u>: $\|(R-R_n)f\| \leq \|(R-R_n)^{\frac{1}{2}}\| \, \|(R-R_n)^{\frac{1}{2}}f\|$ and
$\|(R-R_n)^{\frac{1}{2}}f\|^2 = \langle f,(R-R_n)f\rangle \to 0$.

<u>Theorem 7.9</u> . Let A be a self-adjoint operator which is bounded below. Let A_n be a sequence of self-adjoint operators such that

$A \leq A_n$. Assume that there is a subspace $E \subset Q(A_n)$ for all n such that $\langle f, A_n f\rangle \to \langle f, Af\rangle$ for every f in E . If E is dense in $Q(A)$, then $A_n \to A$ in the sense of strong resolvent convergence.

Proof: We may assume without loss of generality that $0 < c \leq A \leq A_n$. Then $A_n^{-1} \leq A^{-1} \leq c^{-1}$ and for f in H and g in E we have

$$|\langle (A^{-1}-A_n^{-1})f, Ag\rangle| = |\langle (A_n-A)A_n^{-1}f, g\rangle|$$

$$\leq \langle (A_n-A)A_n^{-1}f, A_n^{-1}f\rangle^{\frac{1}{2}}\langle (A_n-A)g,g\rangle^{\frac{1}{2}} \leq \langle f, A_n^{-1}f\rangle^{\frac{1}{2}}\langle (A_n-A)g,g\rangle^{\frac{1}{2}} \to 0 .$$

However the $A_n^{-1}f$ are bounded in $Q(A)$, because $\langle A_n^{-1}f, AA_n^{-1}f\rangle \leq \langle f, A_n^{-1}f\rangle$. So since E is dense in $Q(A)$, $\langle (A^{-1}-A_n^{-1})f, Ag\rangle \to 0$ for all g in $Q(A)$, in particular for g in $D(A)$. Hence $A_n^{-1} \to A^{-1}$ weakly. Since $A_n^{-1} \leq A^{-1}$, it follows th that $A_n^{-1} \to A^{-1}$ strongly.

EXAMPLE

Let $H = L^2(\mathbb{R}^n, dx)$ with $n \geq 2$. Let $H_o = -\Delta$ and $U_n \geq 0$ be a sequence of positive L^1 functions which have support in a ball (centered at some fixed point) with radius which goes to zero with n. Let $H = H_o + U_n$. Then $H_n \to H_o$ in the sense of strong resolvent convergence.

To see this, let E be the space of C^∞ functions with compact support in the complement of the center point. If $n \geq 2$, then E is dense in $Q(H_o)$. Since $H_o \leq H_n$ and $H_n \to H_o$ as forms on E the preceding theorem shows that $H_n \to H$ in the sense of strong resolvent convergence.

The drama of this example is that there is no restriction on the magnitude of U_n as n varies, only on the size of its support. Nevertheless functions of H_n converge to the corresponding functions of H_o. In particular the solutions $\exp(-itH_n)$ of the Schrödinger equations converge to $\exp(-itH_o)$.

Theorem 7.10. Let A_n be an increasing sequence of self-adjoint operators which are bounded below, so that $A_n \leq A_{n+1}$. Let A be a self-adjoint operator such that $A_n \leq A$ and such that whenever f is in $\mathcal{Q}(A_n)$ and $\langle f, A_n f\rangle$ is bounded, then f is in $\mathcal{Q}(A)$ and $\langle f, A_n f\rangle \to \langle f, Af\rangle$. Then $A_n \to A$ in the sense of strong resolvent convergence.

Proof: We may assume that $0 < c \leq A_n$. Since the A_n^{-1} are decreasing and bounded below (by A^{-1}), it follows that the A_n^{-1} have a weak limit, hence a strong limit. Since this limit is bounded below by A^{-1}, it must have trivial kernel. Call the limit B^{-1}. Thus $A^{-1} \leq B^{-1} \leq A_n^{-1}$, which implies that $A_n \leq B \leq A$. Since $A_n \to B$ in the sense of strong resolvent convergence, we need only show that $B = A$.

The trick is to notice that by Proposition 7.7 $A_n^{-1} \to B^{-1}$ strongly implies that $A_n^{-\frac{1}{2}} \to B^{-\frac{1}{2}}$ strongly. Consider the space $\mathcal{Q} = \mathcal{Q}(B)$. We will show that for g in $\mathcal{Q}$, $A_n^{\frac{1}{2}}g \to A^{\frac{1}{2}}g$ strongly in $\mathcal{H}$.

If g is in $\mathcal{Q}$ then $B^{\frac{1}{2}}g$ is in $\mathcal{H}$. If f is in $\mathcal{Q}$ then $A_n^{\frac{1}{2}}f$ is also in $\mathcal{H}$ (since $A_n \leq B$). Hence $\langle f, (A_n^{\frac{1}{2}} - B^{\frac{1}{2}})g\rangle = \langle A_n^{\frac{1}{2}}f, (B^{-\frac{1}{2}} - A_n^{-\frac{1}{2}})B^{\frac{1}{2}}g\rangle \to 0$. Since the $\|A_n^{\frac{1}{2}}g\| \leq \|Bg\|$ are uniformly bounded, it follows that $A_n^{\frac{1}{2}}g \to B^{\frac{1}{2}}g$ weakly. But weak convergence and no loss of norm implies strong convergence. So $A_n^{\frac{1}{2}}g \to B^{\frac{1}{2}}g$ strongly.

In particular, if g is in $\mathcal{Q}(B)$, $\langle g, A_n g\rangle \to \langle g, Bg\rangle$. It follows by assumption that g is in $\mathcal{Q}(A)$ and $\langle g, Ag\rangle = \langle g, Bg\rangle$. Hence A extends B. But $B \leq A$, so $A = B$.

Now we are almost ready for the main approximation theorem. First we review the basic results on form sums that we have obtained so far.

Let $H_o \geq 0$ be a self-adjoint operator acting in $\mathcal{H}$. Let $U \geq 0$ be another self-adjoint operator acting in $\mathcal{H}$. Then U is said to satisfy <u>condition A</u> if $\mathcal{Q}(H_o) \cap \mathcal{Q}(U)$ is dense in $\mathcal{H}$.

Let W be a self-adjoint operator. Then W is said to be <u>form small</u> if there exists $a < 1$ and b such that $\pm W \leq a(H_o+b)$.

<u>Theorem 7.11</u>. Let $\mathcal{H}$ be a Hilbert space and $H_o \geq 0$ be a self-adjoint operator acting in $\mathcal{H}$. Let U and W be self-adjoint operators acting in $\mathcal{H}$ and set $V = U + W$. Assume that $U \geq 0$ and satisfies condition A. Assume that W is form small (in particular $\mathcal{Q}(W) \supset \mathcal{Q}(H_o)$). Then there exists a self-adjoint operator H whose form is the form of $H_o + V = (H_o+U) + W$. Also $\mathcal{Q}(H) = \mathcal{Q}(H_o) \cap \mathcal{Q}(U)$ and H is bounded below, in fact, $H \geq -b$.

<u>Proof</u>: This theorem summarizes the main conclusions of Theorems 4.1 and 5.2.

If A is a self-adjoint operator, the corresponding <u>truncated operators</u> A_k are defined by $A_k = A$ where $|A| \leq k$, $A_k = 0$ where $|A| > k$.

Approximation Theorem 7.12 . Let $H_o \geq 0$ be a self-adjoint operator acting in $\mathcal{H}$. Let $U \geq 0$ and $W \leq 0$ be self-adjoint operators acting in $\mathcal{H}$. Assume that U satisfies condition A and that W is form small. Let U_k and W_r be the corresponding truncated operators. Set $V = U + W$, $V_k = U_k + W$, and $V_{kr} = U_k + W_r$. Then $H_{kr} = H_o + V_{kr} \rightarrow H_k = H_o + V_k$ as $r \rightarrow \infty$ and $H_k \rightarrow H = H_o + V$ as $k \rightarrow \infty$ in the sense of strong resolvent convergence.

Proof: Since $H_k \leq H_{kr}$, the first result follows from Theorem 7.9 . For the second, note that $H_k \leq H$. Thus we may hope to apply Theorem 7.10 . Consider f in $Q(H_k) = Q(H_o)$ with $\langle f,H_k f\rangle$ bounded. Then $\langle f,U_k f\rangle$ is bounded. Since $\langle f,U_k f\rangle \rightarrow \langle f,Uf\rangle$, it follows that f is in $Q(U)$. Hence f is in $Q(H) = Q(H_o) \cap Q(U)$ and $\langle f,H_k f\rangle \rightarrow \langle f,Hf\rangle$. The theorem indeed applies.

NOTES

The standard reference on strong resolvent convergence and on when convergence of forms implies strong resolvent convergence is Chapter VIII of Kato's book (1966).

Wüst (1973) has given a convergence theorem which doesn't depend on the assumption of semi-boundedness. This is applied to Dirac operators by Schmincke (1973).

Parrot (1969) has given examples which illustrate some of the difficulties with more general definitions of limit.

The example of the Schrödinger operator with a potential with small support is due to Friedman (1972). He shows that if the potentials U_n are positive and the capacity of the support of U_n

approaches zero, then $H_o + U_n$ approaches H_o in the strong resolvent sense. However he gives an example when $n = 3$ to show that this may fail if the positivity condition is dropped. (This does not contradict Proposition 7.6, since E is not dense in $\mathcal{D}(H_o)$ until $n \geq 4$.) A paper of Schonbek (1973) has further discussion.

Part II : OPERATOR DOMAINS

§8 ORDER IN HILBERT SPACE

We have seen that there is not much difficulty in adding positive self-adjoint operators. Now we turn to a more detailed examination of the sum. In particular, we would like to see to what extent we can get information about the operator sum.

The interest in knowing about the operator sum may be illustrated by the following theorem of Trotter. Let H_o and V be self-adjoint operators and consider their operator sum $H_o + V$ on $D(H_o) \cap D(V)$. Let H be the closure of the operator sum. The theorem states that if H is self-adjoint, then $\exp(-itH)f = \lim_{n\to\infty}(\exp(-i\frac{t}{n}H_o)\exp(-i\frac{t}{n}V))^n f$ for all f in $\mathcal{H}$. (In the application to quantum mechanics this formula is closely related to the Feynman path integral.)

One of the main tools will be the partial order relation in a Hilbert space $L^2(M,\mu)$. It is more convenient to discuss order for real functions than for complex functions, so we begin with the reduction to real Hilbert space.

A <u>real Hilbert space</u> is a real vector space with an inner product such that the space is complete. Most of the theory of complex Hilbert spaces carries over to real Hilbert spaces. Notice, however, that there is no polarization identity.

Let $\mathcal{H}$ be a complex Hilbert space. Then $T : \mathcal{H} \to \mathcal{H}$ is a <u>conjugation</u> if T is anti-linear $(T(af+bg) = a^*T(f) + b^*T(g))$, anti-unitary $(\langle Tf,Tg\rangle = \langle f,g\rangle^*)$, and $T^2 = 1$. If f is in $\mathcal{H}$ and $Tf = f$, then f is called <u>real</u>. If A is an operator acting in $\mathcal{W}$ such that f in $\mathcal{D}(A)$ implies that Tf is in $\mathcal{D}(A)$ and TAf = ATf , then A is called <u>real</u>.

The set of real elements in a complex Hilbert space forms a real Hilbert space. If A is real, then A leaves this real Hilbert space invariant. Questions about real operators in a complex Hilbert space may usually be reduced to questions about operators in a real Hilbert space.

In the rest of this section we will take the Hilbert space $\mathcal{H}$ to be a real Hilbert space.

Let $\mathcal{H} = L^2(M,\mu)$. An operator $A : \mathcal{H} \to \mathcal{H}$ is said to be <u>positivity preserving</u> if $f \geq 0$ implies $Af \geq 0$. Since A is linear, if it is positivity preserving it is also order preserving, that is, $f \leq g$ implies $Af \leq Ag$.

Notice that the set of positive functions in L^2 is a closed cone. Thus the property of being positivity preserving is preserved under strong limits. This fact will be used in the proof of the following proposition.

<u>Proposition 8.1</u> . Let $\mathcal{H} = L^2(M,\mu)$. Let $H_o \geq 0$ be a self-adjoint operator acting in $\mathcal{H}$. Assume that $(H_o+c)^{-1}$ is positivity preserving for all $c > 0$. Let $U \geq 0$ and $W \leq 0$ be real functions on M . Assume that U satisfies condition A and W is form small. Set $V = U + W$ and $H = H_o + V$. Then $(H+c)^{-1}$ is positivity preserving for all c sufficiently large.

<u>Proof</u>: Let U_k and W_r be the truncations of U and W , so that $0 \leq U_k \leq k$ and $-r \leq W_r \leq 0$. Let $H_{kr} = H_o + U_k + W_r$. It is easy to verify that $(H_{kr}+c)^{-1}$ is positivity preserving for c

sufficiently large. In fact, we may expand this as
$(H_{kr}+c)^{-1} = (H_o+c+k)^{-1}\sum_{n=0}^{\infty}[(k-U_k-W_r)(H_o+c+k)^{-1}]^n$, and each term is positivity preserving.

If $-b$ is less than the lower bound of $H_o + W \leq H_o + W_r \leq H_{kr}$, then we may take $c > b$ and expand
$(H_{kr}+b)^{-1} = (H_{kr}+c)^{-1}\sum_{n=0}^{\infty}[(c-b)(H_{kr}+c)^{-1}]^n$. Hence $(H_{kr}+b)^{-1}$ is positivity preserving for all such b . The conclusion thus follows from the approximation theorem (Theorem 7.12).

The following proposition gives a way of verifying condition A . In fact, the conclusion is stronger than condition A .

Proposition 8.2 . Let $\mathcal{H} = L^2(M,\mu)$. Let e be an element of $\mathcal{H}$ which is strictly positive almost everywhere. Let $H_o \geq 0$ be a self-adjoint operator acting in $\mathcal{H}$ such that $H_o e = 0$. Assume that $(H_o+c)^{-1}$ is positivity preserving for all $c > 0$. Let $U \geq 0$ be a function such that $\langle e,Ue\rangle < \infty$. Then $\mathcal{Q}(H_o) \cap \mathcal{Q}(U)$ is dense in $\mathcal{Q}(H_o)$.

Proof: We will see in fact that $\mathcal{D}(H_o) \cap \mathcal{Q}(U)$ is dense in $\mathcal{D}(H_o)$. Since $\mathcal{D}(H_o)$ is dense in $\mathcal{Q}(H_o)$, this implies the conclusion of the theorem.

Let $L^\infty(e) = \{f \text{ in } L^2 : \pm f \leq te \text{ for some } t\}$. It is easy to see that $L^\infty(e)$ is dense in $\mathcal{H}$. In fact, if $h \perp L^\infty(e)$, then $h \perp e\,\mathrm{sign}(h)$, so $|h| \perp e$. Since $e > 0$ almost everywhere, this implies that $h = 0$.

Next observe that $(H_o+c)^{-1}$ leaves $L^\infty(e)$ invariant. In fact, if $\pm f \leq te$, then $\pm(H_o+c)^{-1}f \leq t(H_o+c)^{-1}e = tc^{-1}e$. If we let $C = (H_o+c)^{-1}L^\infty(e)$, it follows that $C \subset \mathcal{D}(H_o) \cap L^\infty(e)$. However since U is in $L^1(M,e^2\mu)$, $L^\infty(e) \subset \mathcal{Q}(U)$, and so $C \subset \mathcal{D}(H_o) \cap \mathcal{Q}(U)$. Thus it is sufficient to show that C is dense in $\mathcal{D}(H_o)$. But $L^\infty(e)$ is dense in $\mathcal{H}$, so $C = (H_o+c)^{-1}L^\infty(e)$ is dense in $\mathcal{D}(H_o) = (H_o+c)^{-1}\mathcal{H}$.

__Theorem 8.3__ . Let $\mathcal{H} = L^2(M,\mu)$. Let e be an element of $\mathcal{H}$ which is strictly positive almost everywhere. Let $H_o \geq 0$ be a self-adjoint operator acting in $\mathcal{H}$ such that $H_o e = 0$. Assume that $(H_o+c)^{-1}$ is positivity preserving for all $c > 0$. Let $U \geq 0$ be a real function in $L^2(M,e^2\mu)$. Let $H = H_o + U$. Then H is essentially self-adjoint on $\mathcal{D}(H_o) \cap \mathcal{D}(U)$.

__Proof__: Let $L^\infty(e) = \{f$ in $L^2 : \pm f \leq te$ for some $t\}$. As before, $L^\infty(e)$ is dense in $\mathcal{H}$.

First we show that the form sum $H = H_o + U$ is a self-adjoint operator. In fact, $\langle e,Ue\rangle \leq \|e\| \, \|Ue\| < \infty$, so Proposition 8.2 applies. Hence $\mathcal{Q}(H_o) \cap \mathcal{Q}(U)$ is dense in $\mathcal{Q}(H_o)$, which in turn is dense in $\mathcal{H}$.

Now we prove that the restriction of H to $\mathcal{D}(H_o) \cap \mathcal{D}(U)$ is essentially self-adjoint.

We know that $(H+c)^{-1}$ is positivity preserving. We now show that $(H+c)^{-1}e \leq c^{-1}e$. In fact, if we set $H = H_o + U_k$, where $0 \leq U_k \leq k$, we have
$(H_k+c)^{-1}e = (H_o+c)^{-1}e - (H_o+c)^{-1}U_k(H_k+c)^{-1}e \leq (H_o+c)^{-1}e = c^{-1}e$. If we let $k \to \infty$ we obtain $(H+c)^{-1}e \leq c^{-1}e$.

It follows that $(H+c)^{-1}$ leaves $L^\infty(e)$ invariant. In fact, if $\pm f \leq te$, then $\pm(H+c)^{-1}f \leq t(H_o+c)^{-1}e \leq tc^{-1}e$. Let $E = (H+c)^{-1}L^\infty(e)$. Then $E \subset \mathcal{D}(H) \cap L^\infty(e) \subset \mathcal{D}(H) \cap \mathcal{D}(U)$.

We will see that $\mathcal{E} \subset \mathcal{D}(H_o) \cap \mathcal{D}(U)$ and that $\mathcal{E}$ is dense in $\mathcal{D}(H)$. If g is in $\mathcal{E}$, then Hg and Ug are in $\mathcal{H}$. But then $f \mapsto \langle f, H_o g\rangle = \langle f, Hg\rangle - \langle f, Ug\rangle$ is continuous, so f is in $\mathcal{D}(H_o)$. Thus $\mathcal{E}$ is also contained in $\mathcal{D}(H_o)$.

Finally we show that $\mathcal{E}$ is dense in $\mathcal{D}(H)$. But $L^\infty(e)$ is dense in $\mathcal{H}$, and so $\mathcal{E} = (H+c)^{-1}L^\infty(e)$ is dense in $\mathcal{D}(H) = (H+c)^{-1}\mathcal{H}$.

EXAMPLE

Let $\mathcal{H} = L^2(\mathbb{R}^n, dx)$. Define $T : \mathcal{H} \to \mathcal{H}$ by $Tf(x) = f(x)^*$. Then T is a conjugation and the real elements defined by T are the real functions. The corresponding operator $\hat{T} = FTF^{-1}$ in the Fourier transform representation is $\hat{T}\hat{f}(k) = \hat{f}(-k)^*$. Thus the effect of $\hat{T}$ is to reverse momenta.

The operator $-\Delta$ is clearly real with respect to T. Notice that this is a reflection of the fact that the kinetic energy k^2 is even in the momentum k.

Let $H_o = -\Delta$. We will see that $(H_o+c)^{-1}$ is positivity preserving for $c > 0$. It is enough to show that this is true on a dense set, such as $\mathcal{S}$.

Assume f is in $\mathcal{S}$ with $f \geq 0$ and $g = (H_o+c)^{-1}f$ is negative somewhere. Since g is also in $\mathcal{S}$, g must take its minimum value at some point x. But then $f(x) = (-\Delta+c)g(x) \leq cg(x) < 0$, which is a contradiction.

Let $\mathcal{H} = L^2(\mathbb{R}^n, dx)$. Let $H_o = -\Delta$ and $U \geq 0$ be a function on $\mathbb{R}^n$. Assume that U is in $L^2(\mathbb{R}^n, \exp(-2a|x|)dx)$ for some $a < \infty$.

Then $H = H_o + U$ is essentially self-adjoint on $\mathcal{D}(H_o) \cap \mathcal{D}(U)$. This may be proved by means of the following trick.

Let Y be a real function on $\mathbb{R}^n$ which is constant on a ball centred at the origin and is zero elsewhere. Then $-\Delta + Y$ is self-adjoint with the same domain as $-\Delta$, since Y is bounded. By a suitable choice of Y we may arrange that $-\Delta + Y$ has a strictly negative eigenvalue. We may choose it as negative as we please. Let $-c^2$ (where $c > 0$) be the most negative eigenvalue of $-\Delta + Y$ and set $H_o^1 = -\Delta + Y + c^2$. Then $H_o^1 \geq 0$ and zero is an eigenvalue of H_o^1 . It is sufficient to prove the theorem with H_o^1 in place of $-\Delta$.

Let e be the eigenfunction of H_o^1 with $H_o^1 e = 0$. We may choose the sign of e so that $e(x) > 0$ for all x in $\mathbb{R}^n$.

We now use our freedom to choose c arbitrarily large to require that $c > a$. This implies that e satisfies an estimate $e(x) \leq k \exp(-a|x|)$. Hence $\int U(x)^2 e(x)^2 dx \leq k^2 \int U(x)^2 \exp(-2a|x|)dx < \infty$, so U is in $L^2(\mathbb{R}^n, e(x)^2 dx)$ and Theorem 8.3 is applicable.

NOTES

A proof of the theorem of Trotter mentioned in the beginning may be found in §8 of the notes by Nelson (1969).

The essential self-adjointness theorem (Theorem 8.3) is based on the fact that there is an auxiliary Banach space that is left invariant. This type of theorem arose in quantum field theory in the work of Rosen (1970). An abstract formulation appeared in an important paper of Segal (1970). The theory was extended by Simon and Hoegh-Krohn (1972) in their survey of hypercontractive semigroups.

The theory presented here is essentially the specialization of this work to the case when it is sufficient to deal with contractive semi-groups. This specialization was applied in quantum field theory by Hoegh-Krohn (1971). It was also studied by Semenov (1972).

The first application of these ideas to Schrödinger operators was by Simon (1973a). The trick used in the example is due to Faris (1972b), (1973).

§9 ORDER AND ESTIMATES

In order to obtain information about operator sums it would be convenient to be able to pass from one second order estimate to another. There are inequalities which allow this; for example it is always true that if A and B are self-adjoint then $(A+B)^2 \leq 2(A^2+B^2)$. However even when A and B are positive it is not so easy to estimate $A^2 + B^2$ in terms of $(A+B)^2$. The difficulty is that the cross terms $AB + BA$ need not be positive.

In this section we see how this difficulty arises and how it may be avoided in certain cases by use of order properties. The main result (Theorem 9.2) is illustrated by an example from quantum field theory.

We write $C^\infty(A)$ for the space of all vectors f such that $A^n f$ is in $\mathcal{H}$ for all $n = 0,1,2,3,\ldots$. The following theorem shows that an estimate of the cross terms on $C^\infty(A)$ may give information about the domains. The point is that it is not necessary to identify a dense subspace of $\mathcal{D}(A+B)$.

Theorem 9.1 . Let $A \geq 0$ and $B \geq 0$ be self-adjoint operators. Assume that $C^\infty(A) \subset \mathcal{D}(B)$ and is dense in $\mathcal{D}(B)$. Assume also that $AB + BA \geq -a - bA^2$ for some a and $b < 1$ as forms on $C^\infty(A)$. Then $\mathcal{D}(A+B) = \mathcal{D}(A) \cap \mathcal{D}(B)$ and there is an estimate $A^2 + B^2 \leq c(A+B)^2 + d$.

Proof: We need only the estimate which shows that $\mathcal{D}(A+B) \subset \mathcal{D}(A) \cap \mathcal{D}(B)$. Since ABA is positive, we have $AB(1+A/n) + (1+A/n)BA \geq - a(1+A/n)^2 - bA^2$. Let $A_n = A(1+A/n)^{-1}$. It follows that $A_nB + BA_n \geq - a - bA_n^2$ as forms on $C^\infty(A)$. Since

A_n is bounded and $C^\infty(A)$ is dense in $\mathcal{D}(B)$, the same form estimate holds on $\mathcal{D}(B)$. In other words, $A_n^2 + B^2 \leq k(A_n+B+1)^2$ as forms on $\mathcal{D}(B)$.

Since $A_n + B$ increases to the form sum $A + B$, $(A_n+B+1)^{-1}h \to (A+B+1)^{-1}h$ for all h in $\mathcal{H}$, by Theorem 7.10. Let u be in $\mathcal{D}(A+B)$. Then $u = (A+B+1)^{-1}h$ for some h in $\mathcal{H}$. Set $u_n = (A_n+B+1)^{-1}h$. Then u_n is in $\mathcal{D}(B)$ and $\|A_n u_n\|^2$ and $\|Bu_n\|^2$ are bounded by $k\|h\|^2$. It follows from Fatou's lemma that u is in $\mathcal{D}(A) \cap \mathcal{D}(B)$ and $\|Au\|^2 + \|Bu\|^2 \leq k\|h\|^2 = k\|(A+B+1)u\|^2$.

Theorem 9.2. Let $\mathcal{H} = L^2(M,\mu)$. Let $H_o \geq 0$ be a self-adjoint operator acting in $\mathcal{H}$. Assume that $(H_o+c)^{-1}$ is positivity preserving for all $c > 0$. Let $U \geq 0$ be a real measurable function on M such that $\mathcal{Q}(H_o) \cap \mathcal{Q}(U)$ is dense in $\mathcal{H}$. Let $H = H_o + U$. Let W be a real measurable function on M such that for some a and $b > 0$, $W^2 \leq a^2(H_o+b)^2$. Then $W^2 \leq a^2(H+b)^2$.

Proof: Let U_k be equal to U where $U \leq k$, 0 otherwise, and set $H_k = H_o + U_k$. Then $(H_k+c)^{-1}$ is positivity preserving and so if $f \geq 0$ we have
$0 \leq (H_k+c)^{-1}f = (H_o+c)^{-1}f - (H_o+c)^{-1}U_k(H_k+c)^{-1}f \leq (H_o+c)^{-1}f$. Let $k \to \infty$. We see that $0 \leq (H+c)^{-1}f \leq (H_o+c)^{-1}f$. Hence
$0 \leq |W|(H+c)^{-1}f \leq |W|(H_o+c)^{-1}f$.

Now let f be arbitrary. Since
$\pm|W|(H+c)^{-1}f \leq |W|(H+c)^{-1}|f| \leq |W|(H_o+c)^{-1}|f|$, we see that
$\|W(H+c)^{-1}f\| \leq \|W(H_o+c)^{-1}|f|\| \leq \|W(H_o+c)^{-1}\| \, \||f|\| \leq \|W(H_o+c)^{-1}\| \, \|f\|$.
Hence $\|W(H+c)^{-1}\| \leq \|W(H_o+c)^{-1}\|$.

EXAMPLES

Let $\mathcal{H} = L^2(\mathbb{R}^n, dx)$ and $H_o = -\Delta$. Let $U \geq 0$ be a function on $\mathbb{R}^n$ such that $H_o + U$ is essentially self-adjoint on $\mathcal{D}(H_o) \cap \mathcal{D}(U)$. Let W be a real function on $\mathbb{R}^n$ such that $W^2 \leq a^2(H_o+b)^2$ with $a^2 < 1$. Thus W is a small operator perturbation of H_o. Let $V = U + W$. Then $H_o + V$ is essentially self-adjoint on $\mathcal{D}(H_o) \cap \mathcal{D}(V)$.

In fact, by Theorem 9.2 $W^2 \leq a^2(H_o+U+b)^2$, so W is also a small operator perturbation of $H_o + U$. Hence $H_o + V = H_o + U + W$ is essentially self-adjoint on

$\mathcal{D}(H_o) \cap \mathcal{D}(U) = \mathcal{D}(H_o) \cap \mathcal{D}(W) \cap \mathcal{D}(U) \subset \mathcal{D}(H_o) \cap \mathcal{D}(V)$.

The rest of this section is devoted to an example from quantum field theory. The basic idea is to treat fields as quantum mechanical observables and to try to write quantum mechanical equations of motion in terms of the fields. The difficulty is that the requirements of relativistic invariance force a very singular behavior of the interaction terms involving the fields.

These difficulties are much less severe in one and two space dimensions than in three space dimensions. The most progress so far has been in constructing solutions in the one space dimensional case. In this case it has been possible to construct a relativistic theory. A preliminary step in this construction is the construction of a theory on a finite interval, essentially the quantum theory of a non-linear vibrating string.

We begin with the construction of the field. Consider a periodic interval of length L. Let Γ be the set of all integer multiples of $(2\pi/L)$ and let Γ_+ be the set of strictly positive integer multiples of $(2\pi/L)$. The functions $(1/L)^{\frac{1}{2}}$ and $(2/L)^{\frac{1}{2}}\cos(kx)$ and $(2/L)^{\frac{1}{2}}\sin(kx)$, where k ranges over Γ_+, form an orthonormal basis for the functions on the interval. If u is a real function, it has a Fourier series expansion $u(x) = (1/L)^{\frac{1}{2}}a_o + (2/L)^{\frac{1}{2}}\Sigma_k a_k\cos(kx) + b_k\sin(kx)$ with real coefficients.

Let M be the space of all sequences of real numbers. Let the sequences be labeled in the same way as the coefficients of a Fourier series, so that a typical sequence is $\{x_o,x_k,y_k\}$ where k ranges over Γ_+.

Let ν be the product measure on M which is $\pi^{-\frac{1}{2}}\exp(-x^2)dx$ or $\pi^{-\frac{1}{2}}\exp(-y^2)dy$ on each coordinate. The Hilbert space describing the state of the field is $H = L^2(M,\nu)$. Each of the coordinate functions x_o, x_k, or y_k is a multiplication operator in H.

Let m be a positive real number and set $\mu(k) = (k^2+m^2)^{\frac{1}{2}}$. The field is defined by the Fourier series expansion $\phi(x) = (1/L)^{\frac{1}{2}}x_o\mu(0)^{-\frac{1}{2}} + (2/L)^{\frac{1}{2}}\Sigma_k(x_k\cos kx+y_k\sin kx)\mu(k)^{-\frac{1}{2}}$. Thus the Fourier coefficients in its expansion are multiplication operators.

For each x, $\phi(x)$ is a densely defined Hermitian form. In fact, let f and g be functions in $L^2(M,\nu)$ which depend only on finitely many coordinates. Then the matrix elements $\langle f,x_k g\rangle$ and $\langle f,y_k g\rangle$ are zero for k sufficiently large. Hence the sum defining $\langle f,\phi(x)g\rangle$ is a finite sum.

While the field $\phi(x)$ at a point x is not an operator, its average $\int_0^L \phi(x)u(x)dx$ with a sufficiently nice function u is a multiplication operator. This may be seen from the Fourier series expansions of ϕ and u. In fact,

$\int_0^L \phi(x)u(x)dx = a_o x_o \mu(0)^{-\frac{1}{2}} + \Sigma_k (a_k x_k + b_k y_k)\mu(k)^{-\frac{1}{2}}$. The functions x_k and y_k are orthogonal in $L^2(M,\nu)$ and all have the same norm. So if $\Sigma_k (a_k^2 + b_k^2)\mu(k)^{-1} < \infty$, the series expression for the averaged field converges in L^2 to a measurable function.

The operators corresponding to the time derivatives of the Fourier coefficients of the field are $p_{xk} = \frac{1}{i}\left(\frac{\partial}{\partial x_k} - x_k\right)$ and $p_{yk} = \frac{1}{i}\left(\frac{\partial}{\partial y_k} - y_k\right)$. The time derivative field is defined as $\pi(x) = (1/L)^{\frac{1}{2}} p_o \mu(0)^{\frac{1}{2}} + (2/L)^{\frac{1}{2}} \Sigma_k (p_{xk}\cos(kx) + p_{yk}\sin(kx)\mu(k))^{\frac{1}{2}}$. Again this may be interpreted as a densely defined Hermitian form.

The energy for the linear field is given by the formal expression

$$H_o = \tfrac{1}{2}\int_0^L \pi(x)^2 + \nabla\phi(x)^2 + m^2\phi(x)^2 dx - \tfrac{1}{2}\,\Sigma_k \mu(k) .$$

This may be interpreted in the Fourier series representation as a limit of partial sums, and evaluated as

$$H_o = \mu(0)\left[-\frac{1}{2}\,\frac{\partial^2}{\partial x_o^2} + x_o\,\frac{\partial}{\partial x_o}\right] + \Sigma_k \mu(k)\left[-\frac{1}{2}\,\frac{\partial^2}{\partial x_k^2} + x_k\,\frac{\partial}{\partial x_k} - \frac{1}{2}\,\frac{\partial^2}{\partial y_k^2} - y_k\,\frac{\partial}{\partial y_k}\right] .$$

Similarly, the momentum carried by the field in the x direction is $P = -\int_0^L \pi(x)\nabla\phi(x)dx = \sum_k k\frac{1}{i}\left(x_k\,\frac{\partial}{\partial y_k} - y_k\,\frac{\partial}{\partial x_k}\right)$. We now examine these expressions to see what kind of objects they represent.

First consider the operator $N = -\frac{1}{2}\,\frac{\partial^2}{\partial x^2} + x\frac{\partial}{\partial x}$ acting in $L^2(\mathbb{R},\pi^{-\frac{1}{2}}\exp(-x^2)dx)$. This operator factors as $N = a^*a$, where $a^* = 2^{-\frac{1}{2}}(2x - \frac{\partial}{\partial x})$ and $a = 2^{-\frac{1}{2}}\frac{\partial}{\partial x}$. The operators a and a^* satisfy the commutation relation $aa^* = a^*a + 1$ and $a\,1 = 0$. Hence the polynomials $(a^*)^n 1 = 2^{-\frac{n}{2}}(2x - \frac{\partial}{\partial x})^n 1$ are eigenvectors of N with eigenvalues $n = 0,1,2,3,\ldots$. It is known that these Hermite polynomials form a basis for $L^2(\mathbb{R},\pi^{-\frac{1}{2}}\exp(-x^2)dx)$, so N is a self-adjoint operator and its spectrum consists of the positive integers.

Next let us consider the case of two factors. The Hilbert space is $L^2(R^2, \pi^{-1}\exp(-(x^2+y^2))dxdy)$. The energy operator may be written as $-\frac{1}{2}\frac{\partial^2}{\partial x^2} + x\frac{\partial}{\partial x} - \frac{1}{2}\frac{\partial^2}{\partial y^2} + y\frac{\partial}{\partial y}$ and the momentum operator is $\frac{1}{i}\left(x\frac{\partial}{\partial y} - y\frac{\partial}{\partial x}\right) = \frac{1}{i}\frac{\partial}{\partial\theta}$.

It is useful to write these operators in complex form. Let $z = x + iy$ and $z^* = x - iy$ and let $\frac{\partial}{\partial z} = \frac{1}{2}\left(\frac{\partial}{\partial x} - i\frac{\partial}{\partial y}\right)$ and $\frac{\partial}{\partial z^*} = \frac{1}{2}\left(\frac{\partial}{\partial x} + i\frac{\partial}{\partial y}\right)$. Define $N_+ = -\frac{\partial}{\partial z^*}\frac{\partial}{\partial z} + z\frac{\partial}{\partial z}$ and $N_- = -\frac{\partial}{\partial z}\frac{\partial}{\partial z^*} + z^*\frac{\partial}{\partial z^*}$. Then the energy operator is $N_+ + N_-$ and the momentum operator is $N_+ - N_-$. The operators N_+ and N_- are the operators for the number of particles moving in the $+x$ and $-x$ direction.

Let us now examine N_+ and N_- in more detail. They factor as $N_+ = a_+^* a_+$ and $N_- = a_-^* a_-$, where $a_+^* = (z - \frac{\partial}{\partial z^*})$, $a_+ = \frac{\partial}{\partial z}$, $a_-^* = (z^* - \frac{\partial}{\partial z})$, and $a_- = \frac{\partial}{\partial z^*}$. The vectors $(a_+^*)^m (a_-^*)^n 1$ form a basis for $L^2(\mathbb{R}^2, \pi^{-1}\exp(-z^*z)dxdy)$ consisting of eigenvectors of N_+ and N_- . The corresponding eigenvalues of N_+ and N_- are the positive integers m and n .

Now let us return to the entire field. The energy may be written as $H_o = \mu(0)N_o + \Sigma_k \mu(k)(N_{+k} + N_{-k})$ and the momentum as $P = \Sigma_k k(N_{+k} - N_{-k})$. If we sum over all k in Γ these expressions simplify to $H = \Sigma_k \mu(k) N_k$ and $P = \Sigma_k k N_k$. The operators $N_k = a_k^* a_k$ with k in Γ have spectrum $0,1,2,3,\ldots$ and are interpreted as the number of particles with momentum k . The energy of a particle is related to the momentum by the relativistic formula $\mu(k) = \sqrt{k^2+m^2}$.

It follows from the above analysis that the vectors $(a^*_{k_1})^{n_1}\cdots(a^*_{k_j})^{n_j}1$ form a basis for $L^2(M,\nu)$ consisting of eigenvectors of H_o and P. Thus in particular H_o and P are self-adjoint operators. The total energy and momentum may be thought of either as total field energy and momentum or total particle energy and momentum. Of course the field observables $\int\phi(x)u(x)dx$ and the particle number observables N_k do not commute.

The non-linear interaction that has been most studied in this type of model is a certain function of the field. As long as $L < \infty$ it is an operator of multiplication by a real function V such that V is in $L^2(M,\nu)$ and $\exp(-V)$ is in $L^p(M,\nu)$ for all $p < \infty$.

The operator H_o is positive, self-adjoint and satisfies $H_o 1 = 0$. It is not difficult to see that $(H_o+c)^{-1}$ is positivity preserving for all $c > 0$.

The crucial ingredient in the study of this model is an infinite dimensional analog of the Sobolev inequality. This reads

$$\int|f|^2\log|f|d\nu \leqslant \frac{1}{m}\langle f,H_of\rangle + \|f\|_2^2\log\|f\|_2 .$$

Write $H = H_o + V = H_o + U + W$ where $U \geqslant 0$ and $W \leqslant 0$ are the positive and negative parts of V. Then U is in $L^2(M,\nu)$ and $\exp(-W)$ is in $L^p(M,\nu)$ for all $p < \infty$. Insert $g = -W/am$ in the inequality $\int g|f|^2d\nu \leqslant \int|f|^2\log|f|d\nu - \|f\|_2^2\log\|f\|_2 + \|f\|_2^2\log\|\exp(g)\|_2$. It follows that

$$-W \leqslant aH_o + \log\|\exp(-W)\|_{\frac{2}{am}} .$$

Thus if $m > 0$, then W is form small. From Proposition 8.2 it is clear that U satisfies Condition A. Hence by Theorem 7.11 $H = H_o + V$ is a self-adjoint operator.

One can also pose the question of whether $H = H_o + V$ is essentially self-adjoint when restricted to the intersection of the domains. That this is true for $H_o + U$ follows from Theorem 8.3 . It has been shown that W is actually operator small with respect to H_o , so Theorem 9.2 implies that it is also true for $H = H_o + U + W = H_o + V$.

NOTES

Glimm and Jaffe (1969) gave an abstract theorem on self-adjointness of the operator sum based on the analysis of cross terms in $(T^2+B)^2$. The idea is to use the double commutator identity $T^2B + BT^2 = 2TBT + [T,[T,B]]$. If $B \geq 0$, then $TBT \geq 0$, so the problem is to estimate the double commutator. There is an interesting extension of the Glimm-Jaffe technique in a paper by Konrady (1971). The cross terms in the case of differential operators have been studied in a series of papers by Everitt and Giertz (1972) ( ). The abstract Theorem 9.1 presented here is adapted from an recent paper by Okazawa (1973). There is related material in a paper by Gustafson and Rejto (1973).

A first attempt to use arguments based on order to allow negative perturbations may be found in Faris (1972b). The basic idea of Theorem 9.2 is due to Davies (1973). Faris (1973) showed that this technique leads to second order estimates. The relation to the maximum principle was pointed out by H. Weinberger. (See the Protter-Weinberger (1967) book on the maximum principle.) In the application to quantum fields this represents an alternative to, or at least a variant of the approach based on hypercontractive semigroups mentioned before.

The origin of these techniques is closely connected with the development of quantum field theory. The model described here was treated rigorously by Nelson (1966). He obtained estimates which showed that the negative part of the interaction is form small. The part of the estimates involving H_o are related to a hypercontractive inequality and to the logarithmic inequality used here. The development of the hypercontractive inequality is surveyed by Nelson (1973); the fact that it is equivalent to logarithmic inequalities was discovered by Gross ( ). His paper also treats the applications.

The construction of a fully relativistic model was undertaken by Glimm and Jaffe. They have described their earlier work in several surveys (1971) (1972). The recent work on this subject tends to emphasize space-time symmetry and give less empasis to the Hamiltonian. There are surveys of some of this in the notes edited by Velo and Wightman (1973).

§10 EIGENVALUES

The conditions for the definition of the form sum of two self-adjoint operators are too general to allow us to say much about the spectrum. However a somewhat stronger but still not very restrictive condition implies certain properties of the eigenvalues. This condition guarantees continuity of the eigenvalues in the coupling constant. Also in the presence of an order structure in the Hilbert space it allows an application of the Perron-Frobenius theorem to the form sum. This gives a method for proving that the lowest eigenvalue has multiplicity one (the ground state is unique).

Let $H_o \geq 0$ be a self-adjoint operator acting in the Hilbert space $\mathcal{H}$. Let $U \geq 0$ be another self-adjoint operator acting in $\mathcal{H}$. Then U is said to satisfy condition B if $\mathcal{Q}(H_o) \cap \mathcal{Q}(U)$ is dense in $\mathcal{Q}(H_o)$.

Notice that condition B differs from condition A only in one respect. In condition A $\mathcal{Q}(H_o) \cap \mathcal{Q}(U)$ is required to be dense in $\mathcal{H}$. In condition B it is required to be dense in $\mathcal{Q}(H_o)$ (with respect to the Hilbert space norm of $\mathcal{Q}(H_o)$). Since $\mathcal{Q}(H_o)$ is dense in $\mathcal{H}$, condition B implies condition A.

Let H_o and W be self-adjoint operators and set $H_\lambda = H_o + \lambda W$, where λ is real. Assume that $H_\lambda f_\lambda = E_\lambda f_\lambda$, where E_λ is real and $\langle f_\lambda, f_\lambda \rangle = 1$. Then $E_\lambda = \langle f_\lambda, H_\lambda f_\lambda \rangle$, so

$$\frac{d}{d\lambda}E_\lambda = \langle \frac{d}{d\lambda}f_\lambda, H_\lambda f_\lambda \rangle + \langle f_\lambda, H_\lambda \frac{d}{d\lambda}f_\lambda \rangle + \langle f_\lambda, W f_\lambda \rangle$$

$$= E_\lambda \frac{d}{d\lambda}\langle f_\lambda, f_\lambda \rangle + \langle f_\lambda, W f_\lambda \rangle = \langle f_\lambda, W f_\lambda \rangle .$$

This makes it plausible that if $\langle f_\lambda, W f_\lambda \rangle$ is finite at any particular value of λ, then E_λ is differentiable there.

If W is a small form perturbation of H_o, then f_λ is in $\mathcal{Q}(H_\lambda) = \mathcal{Q}(H_o) \subset \mathcal{Q}(W)$, so $\langle f_\lambda, W f_\lambda \rangle$ is indeed finite. So we expect that in this case the eigenvalue E_λ depends differentiably on λ.

The situation is not so nice for large perturbations; the eigenvalues need not even be continuous in the coupling constant. However the next theorem shows that condition B implies continuity.

First we need a definition of the <u>increasing sequence of eigenvalues</u> at the bottom of the spectrum. Let H be a self-adjoint operator which is bounded below. Let $E_n = \inf\{c : H \leq c$ has dimension $\geq n\}$. The sequence $E_1 \leq E_2 \leq E_3 \leq \ldots$ is increasing. It is always true that $H < E_n$ has dimension $< n$. If E_n is not a limit point of the spectrum of H, then $H \leq E_n$ has dimension $\geq n$ and E_n is an eigenvalue.

<u>Proposition 10.1</u>. Let H and $\hat{H}$ be two self-adjoint operators which are bounded below. If $H \leq \hat{H}$, then $E_n \leq \hat{E}_n$.

<u>Proof</u>: Let $\varepsilon > 0$. Then $\hat{H} \leq \hat{E}_n + \varepsilon$ has dimension $\geq n$. On the other hand, $H < E_n$ has dimension $< n$. Thus there is a unit vector g in the first space which is orthogonal to the second space. It follows that $E_n \leq \langle g, Hg \rangle \leq \langle g, \hat{H} g \rangle \leq \hat{E}_n + \varepsilon$.

<u>Theorem 10.2</u>. Let $H_o \geq 0$ be a self-adjoint operator. Let $U \geq 0$ be a self-adjoint operator satisfying condition B. Let $\lambda \geq 0$ and set $H_\lambda = H_o + \lambda U$. Then $E_{\lambda n} \to E_{on}$ as $\lambda \to 0$.

<u>Proof</u>: Since $H_o \leq H_\lambda$, $E_{on} \leq E_{\lambda n}$. Let $\varepsilon > 0$. There is a subspace of dimension n on which $H_o \leq E_{on} + \varepsilon$. This subspace is necessarily contained in $\mathcal{Q}(H_o)$. By condition B there is a subspace of dimension n contained in $\mathcal{Q}(H_o) \cap \mathcal{Q}(U)$ on which

$H_o \leq E_{on} + 2\varepsilon$. On the other hand, the subspace on which $H_\lambda < E_{\lambda n}$ has dimension $< n$. Thus we may choose a vector g in $\mathcal{Q}(H_o) \cap \mathcal{Q}(U)$ with $E_{\lambda n} \leq \langle g,H_\lambda g\rangle = \langle g,H_o g\rangle + \lambda\langle g,Ug\rangle \leq E_{on} + 2\varepsilon + \lambda\langle g,Ug\rangle$.

Since $\langle g,Ug\rangle$ is finite, this is less than $E_{on} + 3\varepsilon$ for sufficiently small λ .

For the application of the Perron-Frobenius theorem we need certain facts about invariant subspaces. Let A be a bounded operator acting in the Hilbert space $\mathcal{H}$. Let P be the projection onto some closed subspace. Then A leaves invariant the subspace if and only if $AP = PAP$. The projection onto the orthogonal complement of the subspace is $1 - P$. If A also leaves this invariant, then $PA(1-P) = 0$, so $AP = PAP + PA(1-P) = PA$, that is, A commutes with P . In this case we say that the closed subspace <u>reduces</u> A .

If A is a bounded operator which leaves a closed subspace invariant, its adjoint A^* leaves the orthogonal complement invariant. Thus an invariant closed subspace for a self-adjoint operator actually reduces the operator.

Let $\mathcal{H} = L^2(M,\mu)$. A <u>configuration projection</u> is a projection which is a multiplication operator. The range of such a projection is a <u>configuration subspace</u>.

A configuration projection is multiplication by the indicator function 1_S of some measurable subset $S \subset M$. The associated configuration subspace may be identified with $L^2(S) \subset L^2(M)$.

Consider a Hilbert space $\mathcal{H} = L^2(M,\mu)$. Let $A : \mathcal{H} \to \mathcal{H}$ be a bounded operator. Then A is said to be <u>indecomposable</u> if it leaves invariant no non-trivial configuration subspace.

The following result is a Hilbert space version of the Perron-Frobenius theorem. Notice that the cone of positive elements may have empty interior.

<u>Theorem 10.3</u> . Let $\mathcal{H} = L^2(M,\mu)$. Let $A : \mathcal{H} \to \mathcal{H}$ be a bounded self-adjoint operator. Assume that $A \leq a$ where a is an eigenvalue of A . Assume also that A is positivity preserving. Then A is indecomposable if and only if the eigenvalue a has multiplicity one and the corresponding eigenspace is spanned by a function u which is strictly positive almost everywhere.

<u>Proof</u>: Assume that A is indecomposable and let u be a unit vector with $Au = au$. Since A is positivity preserving, it also preserves the real functions, so we may assume that u is real. Then $a = \langle u,Au\rangle \leq \langle |u|,A|u|\rangle$, so $A|u| = a|u|$. Let $u_\pm$ be the positive and negative parts of u . It follows that $Au_\pm = au_\pm$.

Let $S_\pm$ be the set where $u_\pm$ vanishes. If $f \geq 0$ is in $L^2(S_\pm)$, then $\langle Af,u_\pm\rangle = \langle f,Au_\pm\rangle = a\langle f,u_\pm\rangle = 0$, so $Af \geq 0$ is also in $L^2(S_\pm)$. Since A is indecomposable, either $u_+ = 0$ or $u_- = 0$. Hence either u or $-u$ is strictly positive almost everywhere.

This reasoning applies to any real eigenvector. But no two strictly positive functions can be orthogonal. So a must have multiplicity one.

Assume now on the other hand that $Au = au$, where $u > 0$ almost everywhere, and that a has multiplicity one. Let $L^2(S) \subset L^2(M)$ be invariant under A , and let P be the projection onto $L^2(S)$. Since A is self-adjoint, $L^2(S)$ reduces A , and so A commutes with P . Hence $APu = aPu$. Since a has multiplicity one, $Pu = u$ or $Pu = 0$. But since $u > 0$ almost everywhere, this in turn implies $P = 1$ or $P = 0$.

In order to apply this theorem to the resolvents of unbounded operators, it will be useful to consider some facts about projections which commute with self-adjoint operators. If H is an operator and P is a projection, to say that P <u>commutes</u> with H means that f in $\mathcal{D}(H)$ implies that Pf is in $\mathcal{D}(H)$ and $HPf = PHf$. If H is self-adjoint and P commutes with H , then H and P may be simultaneously realized as multiplication operators in some L^2 space, by the spectral theorem for commuting operators. Thus if H is a self-adjoint operator and c is not in the spectrum of H , then P commutes with H if and only if P commutes with $(H+c)^{-1}$. When H is a self-adjoint operator acting in $L^2(M,\mu)$, we say that H is <u>indecomposable</u> if it commutes with no non-trivial configuration projection. Thus a semi-bounded self-adjoint operator H is indecomposable if and only if $A = (H+c)^{-1}$ is indecomposable for sufficiently large c .

<u>Proposition 10.4</u> . Let $\mathcal{H} = L^2(M,\mu)$. Let $H_o \geq 0$ be a self-adjoint operator acting in $\mathcal{H}$. Let U and W be real measurable functions on M with $U \geq 0$ and $W \leq 0$. Assume that U satisfies condition A and W is form small. Set $V = U + W$ and $H = H_o + V$. Let $\hat{H}_o$ be the self-adjoint operator whose form is the closure of the form of

H_o restricted to $Q(H) = Q(H_o) \cap Q(V)$. If $\hat{H}_o$ is indecomposable, then H is indecomposable.

Lemma . Let H be a self-adjoint operator and P be a projection. Then P commutes with H if and only if g in $Q(H)$ implies Pg is in $Q(H)$ and $\langle f,HPg\rangle = \langle Pf,Hg\rangle$ for all f and g in $Q(H)$.

Proof: The only thing worth proving is that the condition on the forms implies P commutes with H . Assume g is in $\mathcal{D}(H)$. Then Hg is in $\mathcal{H}$ and $\langle f,HPg\rangle = \langle Pf,Hg\rangle = \langle f,PHg\rangle$. Hence Pg is in $\mathcal{D}(H)$ and $HPg = PHg\rangle$.

Proof of Proposition: Let P be a configuration projection. If P commutes with H , then $\langle f,HPg\rangle = \langle Pf,Hg\rangle$ for all f and g in $Q(H)$. Since P commutes with the multiplication operator V , it follows that $\langle f,H_oPg\rangle = \langle Pf,H_og\rangle$ for all f and g in $Q(H)$.

In particular, for g in $Q(H)$ we have Pg in $Q(H)$ and

$$\begin{aligned}\langle Pg,H_oPg\rangle &\leq \langle Pg,H_oPg\rangle + \langle(1-P)g,H_o(1-P)g\rangle \\ &= \langle Pg,H_og\rangle + \langle(1-P)g,H_og\rangle \\ &= \langle g,H_og\rangle .\end{aligned}$$

This says that $g \mapsto Pg$ is continuous in the $Q(H_o)$ sense on $Q(H)$. Since by definition $Q(H)$ is dense in $Q(\hat{H}_o)$, it follows that g in $Q(\hat{H}_o)$ implies Pg is in $Q(\hat{H}_o)$ and that P is actually a continuous operator on $Q(\hat{H}_o)$. Thus the relation $\langle f,\hat{H}_oPg\rangle = \langle Pf,\hat{H}_og\rangle$ must in fact be true for all f and g in $Q(\hat{H}_o)$. This implies that P commutes with $\hat{H}_o$.

Theorem 10.5 . Let $\mathcal{H} = L^2(M,\mu)$. Let $H_o \geq 0$ be a self-adjoint operator acting in $\mathcal{H}$. Assume that $(H_o+c)^{-1}$ is positivity preserving for $c > 0$. Let U and W be real measurable functions on M with $U \geq 0$ and $W \leq 0$. Assume that U satisfies condition A and W is form small. Set $V = U + W$ and $H = H_o + V$. Let $\hat{H}_o$ be the self-adjoint operator associated with the closure of the form of H_o restricted to $Q(H) = Q(H_o) \cap Q(V)$. Assume that $\hat{H}_o$ is indecomposable. If $H \geq b$, where b is an eigenvalue of H, then b has multiplicity one and the eigenspace is spanned by a function u which is strictly positive almost everywhere.

<u>Proof</u>: Let $A = (H+c)^{-1}$, where $c > -b$. Then $A \leq a = (b+c)^{-1}$. By Proposition 8.1 A is positivity preserving. Proposition 10.4 shows that A is indecomposable. Hence the eigenvalue a of A has multiplicity one, by Theorem 10.3. But this is equivalent to saying that the eigenvalue b of H has multiplicity one.

If condition B is satisfied, then $\hat{H}_o = H_o$ and it is enough to show that H_o is indecomposable. Sometimes Proposition 8.2 and Theorem 10.4 give a convenient way to establish condition B and indecomposability.

<u>EXAMPLES</u>

Let $\mathcal{H} = L^2(\mathbb{R},dx)$ and $H_o = \frac{1}{2}\left[-\frac{d^2}{dx^2}+x^2\right]$. Then $H_o \geq \frac{1}{2}$ and $H_o e = \frac{1}{2}e$, where $e(x) = \exp(-\frac{1}{2}x^2)$.

Let $U \geq 0$ be a measurable function on $\mathbb{R}$. If U is locally in L^1 on the complement of a closed set of measure zero, then condition A is satisfied. (Thus $H_\lambda = H_o + \lambda U$ is a self-adjoint operator for $\lambda \geq 0$.) However this is not sufficient for condition B. In fact, if $U(x) = (1/x^4)$, then U is locally in L^1 except at the origin. However for all $\lambda > 0$ the lowest eigenvalue of

H_λ is $\geq \frac{3}{2}$. Hence the eigenvalue is not continuous in λ at zero. Also, the lowest eigenvalue of H_λ has multiplicity two. So for either reason condition B must fail.

It follows from Proposition 8.2 that if U is in $L^1(\mathbb{R},\exp(-x^2)dx)$ (that is, $\langle e,Ue\rangle$ is finite), then condition B holds. In fact, it is not difficult to see directly that it suffices that U be locally in L^1 .

Now let us examine the continuity of the eigenvalue in the higher dimensional case. Let K be a compact subset of $\mathbb{R}^n$, where $n \geq 3$. We say that K has <u>capacity zero</u> if whenever v is a tempered distribution with support contained in K and with finite electrostatic energy $\langle v,(-\Delta)^{-1}v\rangle$, then $v = 0$. A surface of codimension 2 (or less) has capacity zero.

Let $\mathcal{H} = L^2(\mathbb{R}^n,dx)$, $n \geq 3$, and $H_o = \frac{1}{2}(-\Delta+x^2)$. Let $U \geq 0$ be a function on $\mathbb{R}^n$. Assume that U is locally integrable on the complement of a compact set K of capacity zero. Then condition B is satisfied. (As a consequence the eigenvalues are continuous in the coupling constant.)

<u>Proof</u>: We must show that the space of C^∞ functions with compact support in the complement K^c of K are dense in $\mathcal{Q}(H_o)$. It is sufficient to show that they are dense in the space of functions with $\langle u,-\Delta u\rangle < \infty$. Let v be a continuous linear functional on this space which vanishes on $C^\infty_{com}(K^c)$. Then v is a distribution of finite electrostatic energy supported on K . Since the capacity of K is zero, $v = 0$. This proves the density.

Next we turn to uniqueness of the ground state in the higher dimensional case. Here the condition is that the perturbation doesn't erect a barrier between different regions.

Let $\mathcal{H} = L^2(\mathbb{R}^n, dx)$. Let $H_o = -\Delta$ and let $U \geq 0$ be a function on $\mathbb{R}^n$ which is locally integrable on the complement of a closed set K of measure zero. Assume that the complement of K is connected. Then the ground state of $H = H_o + U$ (if it exists) is unique.

<u>Proof</u>: This is an application of Theorem 10.5 . (For simplicity we are considering a case where the perturbation is positive.) The only thing we have to check is that $\hat{H}_o$ is indecomposable.

Let S be a measurable subset of $\mathbb{R}^n$. The <u>support</u> of S is the set of all x in $\mathbb{R}^n$ such that for all neighborhoods N of x , $N \cap S$ has strictly positive measure. The support of S is a closed subset of $\mathbb{R}^n$.

Let P be a configuration projection. Then P is multiplication by 1_S , where S is a measurable subset of $\mathbb{R}^n$. Let $\partial S = \operatorname{supp} S \cap \operatorname{supp} S^c$, where S^c is the complement of S . The first assertion is that if P commutes with $\hat{H}_o$, then $\partial S \subset K$.

To see this, consider x not in K . Let u be a function in $C^\infty_{com}(K^c)$ such that $u = 1$ near x . Then u is in $D(\hat{H}_o)$ and so near x the commutativity gives $\Delta 1_S = \Delta 1_S u = 1_S \Delta u = 0$. But this implies that 1_S is continuous near x , so x cannot belong to ∂S .

Now we also use the assumption that K^c is connected. Since $\operatorname{supp} S \cup \operatorname{supp} S^c = \mathbb{R}^n$ and $\partial S = \operatorname{supp} S \cap \operatorname{supp} S^c \subset K$, this implies that $\operatorname{supp} S \subset K$ or $\operatorname{supp} S^c \subset K$. Since K is of measure zero, this implies that $P = 0$ or $P = 1$. This completes the proof.

The theorems on uniqueness of the ground state also apply to the quantum field example. Thus there is no degeneracy - at least not until the limit $L \to \infty$ is taken!

NOTES

The proof of continuity of the eigenvalues in the coupling constant given here is based on a simple variational argument. It provides an alternative to an argument in Kato's book (1966; Chap. VIII, §3). He shows that in the situation of Theorem 7.9 the strong resolvent convergence is enough to imply continuity of the eigenvalues.

The importance of the question of continuity of the eigenvalues may be seen from the case of a large perturbation for which the Taylor series expansion for the eigenvalues is defined. The size of the remainder term depends on the estimates which prove continuity. (There is a discussion of this in a paper by Simon (1971c).)

The example of discontinuity of the eigenvalues is due to Klauder and is discussed further by Simon (1973d).

The application of the Perron-Frobenius theorem in quantum mechanics is due to Glimm and Jaffe (1971). Segal (1971) showed how this type of result could be deduced from an abstract perturbation theorem. The theory was generalized to a von Neumann algebra setting by Gross (1972) and to ordered Hilbert space by Faris (1972a). (Simon (1973e) showed that certain conditions related to indecomposability which are used in this work are actually equivalent.) The proof of the present version (Theorem 10.5) of the perturbation theorem follows another paper of Faris (1972b). However the statement of the theorem is slightly more general. Simon remarked that this allows the example to be formulated with a topological condition for indecomposability, and he gave a proof of indecomposability using Wiener path integrals.

Gross (   ) has given a different sort of criterion for uniqueness of the ground state, based on analytic vector ideas. Gross (1972) has also proved an existence theorem for the ground state. He is able to avoid strong compactness assumptions by the use of order properties.

The notion of capacity used in the example is not completely standard (due to the use of distributions instead of measures). However in the present context it is equivalent to the more usual formulation. (This is discussed in a paper of Deny (1950).)

§11 LOCAL CONDITIONS

The results on operator sums that are obtained in the context of Hilbert space theory all involve global conditions on the perturbation. In order to obtain a result involving only local conditions it is necessary to develop a more specialized theory. In this section the theory of distributions on Euclidean space is applied to give a particularly sharp result for Schrödinger operators.

Consider $C^{\infty}_{com}(\mathbb{R}^n)$, the space of C^{∞} functions on $\mathbb{R}^n$ which have compact support. The elements of this space are called <u>test</u> functions.

Let $K \subset \mathbb{R}^n$ be a compact set and $C^{\infty}(K)$ be the subspace of functions in C^{∞}_{com} which have support in K . A sequence of elements g_n in $C^{\infty}(K)$ is said to converge to g in $C^{\infty}(K)$ if for every p $D^p g_n \to D^p g$ uniformly.

A <u>distribution</u> is a conjugate linear functional on C^{∞}_{com} such that for every compact $K \subset \mathbb{R}^n$, the restriction of the functional to $C^{\infty}(K)$ is continuous. If v is a distribution and g is in C^{∞}_{com} , the value of v on g will be written $\langle g,v \rangle$.

A sequence of distributions v_n is said to <u>converge</u> to a distribution v if $\langle g,v_n \rangle \to \langle g,v \rangle$ for every test function g .

A distribution v is said to be <u>positive</u> if $g \geq 0$ implies $\langle g,v \rangle \geq 0$. This notion of positivity defines a partial order on the space of distributions.

If u is a locally integrable function on $\mathbb{R}^n$, it is integrable over any compact subset $K \subset \mathbb{R}^n$. Thus it defines a distribution by $\langle g,u \rangle = \int g^* u \, dx$, where g is a test function. A basic lemma of distribution theory is that this distribution determines the function (almost everywhere).

If v is a distribution, its <u>derivative</u> $D^p v$ is defined by $\langle g, D^p v\rangle = (-1)^{|p|}\langle D^p g, v\rangle$. (Thus Δv is given by $\langle g, \Delta v\rangle = \langle \Delta g, v\rangle$.)

The space C^∞_{com} is contained in the space S . Recall that a tempered distribution is a continuous conjugate linear functional on S . Every tempered distribution defines a distribution, by restriction. It is a fact that C^∞_{com} is dense in S , so this distribution determines the tempered distribution. Thus the tempered distributions may be identified with certain distributions. It is also a fact that the positive elements of C^∞_{com} are dense in the positive elements of S , so the two possible notions of positivity of a tempered distribution coincide.

There are certainly locally integrable functions that do not determine tempered distributions - for example an exponential function. However the fact that such functions are excluded can actually be an advantage, as we shall now see.

The space O_M is defined as the space of <u>slowly increasing C^∞ functions</u>. (Slowly increasing means that each derivative is bounded at infinity by some polynomial in $|x|$.) The elements of O_M operate on S by multiplication. That is, if ϕ is in O_M and g is in S , then ϕg is also in S .

If ϕ is in O_M , then $\phi(\frac{1}{i}D)$ acts on the space of tempered distributions. In fact $\langle g, \phi(\frac{1}{i}D)v\rangle = \langle \phi^*\hat{g}, \hat{v}\rangle$, so this corresponds to a multiplication operator in the Fourier transform representation. For example $(1-\Delta)^{-1}$ corresponds to $\phi(k) = (1+k^2)^{-1}$, so $(1-\Delta)^{-1}$

leaves the space of tempered distributions invariant. Notice that $1-\Delta$ has no inverse as an operator on the space of distributions.

The inequality needed for the proof of the main theorem is developed in the following two lemmas.

Preparatory Lemma. Fix $\varepsilon > 0$ and set $s(x) = \sqrt{x^2+\varepsilon^2}$. Let u be a real function such that u and Δu are in L^1 locally. Then $\Delta s(u) \geqslant s'(u)\Delta u$.

Proof: If u is smooth we have
$\Delta s(u) = s''(u)(\nabla u)^2 + s'(u)\Delta u \geqslant s'(u)\Delta u$.

In the general case let u^ρ be a sequence of smooth functions such that $u^\rho \to u$ and $\Delta u^\rho \to \Delta u$ locally in L^1. Since $|s(u^\rho)-s(u)| \leqslant |u^\rho - u|$, $s(u^\rho) \to s(u)$ locally in L^1. Hence $\Delta s(u^\rho) \to \Delta s(u)$ as distributions.

Now assume that the u^ρ also converge to u almost everywhere. We will show that $s'(u^\rho)\Delta u^\rho \to s'(u)\Delta u$ locally in L^1. But $s'(u^\rho)\Delta u^\rho - s'(u)\Delta u = s'(u^\rho)(\Delta u^\rho - \Delta u) + (s'(u^\rho)-s'(u))\Delta u$. The first term goes to zero since s' is bounded, while the second term goes to zero by the dominated convergence theorem.

Since inequalities are preserved under distribution limits, the general case follows from the smooth case.

Lemma. Let u be a real function such that u and Δu are in L^1 locally. Then $\Delta|u| \geqslant \operatorname{sign}(u)\Delta u$.

Proof: Let $\varepsilon > 0$ and set $s_\varepsilon(x) = \sqrt{x^2+\varepsilon^2}$. Notice that $s_\varepsilon(x) \to |x|$ uniformly and $s'_\varepsilon(x) \to \operatorname{sign} x$ pointwise as $\varepsilon \to 0$. Hence $\Delta s_\varepsilon(u) \to \Delta|u|$ as distributions and $s'_\varepsilon(u)\Delta u \to \operatorname{sign}(u)\Delta u$

locally in L^1 (by the dominated convergence theorem). The result follows from the previous lemma.

Theorem 11.1 . Assume that $U \geq 0$ and is locally in $L^2(\mathbb{R}^n)$. Then $-\Delta + U$ (on the space of C^∞ functions with compact support) is essentially self-adjoint.

Proof: It is sufficient to show that the image of the C^∞ functions with compact support under $-\Delta + U + 1$ is dense in L^2 .

Let f be a real function in L^2 which is orthogonal to this range. Then $(-\Delta+U+1)f = 0$ in the sense of distributions. Thus $\Delta f = (U+1)f$ is locally in L^1 , so by the lemma,
$\Delta|f| \geq \Delta f \operatorname{sign} f = (U+1)f \operatorname{sign} f = (U+1)|f| \geq |f|$. Hence $(1-\Delta)|f| \leq 0$. But $(1-\Delta)^{-1}$ acts on the space of tempered distributions and preserves positivity, so $|f| \leq 0$, $f = 0$.

NOTES

The classic treatise on distributions is by Schwartz (1966).

The theorem of this section is due to Kato (1972). The technique is applied to local singularities by Simon (1973b) and by Kalf and Walter (1973). (Kalf and Walter (1972) and Schmincke (1972) had earlier results based on partial differential equation methods.) Simon (1973c) has also given an application of the technique to magnetic vector potentials.

If $U \geq 0$ is only locally in L^1 , then one cannot expect that the operator sum $-\Delta + U$ is essentially self-adjoint. But Kato (1974) has shown that his inequality may still be used to obtain information about the domain of the form sum $-\Delta + U$.

§12 COMMUTATORS

Another approach to proving self-adjointness is through commutation properties. If a Hermitian operator almost commutes with a self-adjoint operator, this may force it to be self-adjoint. An elementary result of this nature is presented here.

In addition, an example is given in which the fact that the Hermitian operator has certain commutation properties with respect to a unitary group implies that it is self-adjoint. The result is rather special, but is interesting in that it is not limited to second order differential operators.

Theorem 12.1 . Let H be a Hermitian operator and N a positive self-adjoint operator. Assume that

(i) $\mathcal{D}(N) \subset \mathcal{D}(H)$

and for some constant c and all f in $\mathcal{D}(N)$

(ii) $\pm i\{\langle Hf,Nf\rangle - \langle Nf,Hf\rangle\} \leqslant c\langle f,Nf\rangle$.

Then H is essentially self-adjoint.

Proof: Without loss of generality we may assume that $N \geqslant 1$. We use the fact that H is essentially self-adjoint provided that the range of $H - bi$ is dense whenever $|b|$ is sufficiently large.

Let f be orthogonal to the range of $H - bi$. Then in particular we have from (i) that $\langle f,(H-bi)N^{-1}f\rangle - \langle (H-bi)N^{-1}f,f\rangle = 0$. It follows from this and from (ii) that $\pm 2b\langle f,N^{-1}f\rangle = \pm i\{\langle HN^{-1}f,f\rangle - \langle f,HN^{-1}f\rangle\} \leqslant c\langle f,N^{-1}f\rangle$. If $2|b| > c$, this implies $f = 0$.

EXAMPLE

The following example is a self-adjoint Schrödinger operator which is not bounded below. In order to emphasize the commutation properties we write $p_j = -i\frac{\partial}{\partial x_j}$ and q_k = multiplication by x_k. These operators have the commutation relation $[p_j, q_k] = -i\delta_{jk}$.

Let $\mathcal{H} = L^2(\mathbb{R}^n, dx)$ and $H_o = p^2 = -\Delta$. Let U be a real measurable function on $\mathbb{R}^n$ which is locally in L^2 and satisfies $U(x) \geq -cx^2$ for some constant c. The operator of multiplication by U will be written as $U(q)$. Thus $p^2 + U(q)$ acting on the space C^∞_{com} (of C^∞ functions with compact support) is a Hermitian operator. Let H be its closure. The assertion is that H is self-adjoint.

To prove this, we choose an auxiliary self-adjoint operator $N \geq 0$. The choice we make is $N = p^2 + U(q) + 2cq^2$. Clearly $N \geq p^2 + cq^2 \geq 0$. We know from Theorem 11.1 that C^∞_{com} is dense in $\mathcal{D}(N)$.

We wish to apply Theorem 12.1. In order to verify hypothesis (i) it is sufficient to show that the inequality $\|Hf\|^2 \leq \|Nf\|^2 + b\|f\|^2$ holds for all f in C^∞_{com}. This follows from the double commutator identity

$$\begin{aligned} H^2 &= (N-2cq^2)^2 = N^2 - 2cq^2N - N2cq^2 + 4c^2q^4 \\ &= N^2 - 4c\Sigma_j q_j N q_j + 4cq^4 - 2c\Sigma_j [q_j,[q_j,N]] \\ &= N^2 - 4c\Sigma_j q_j (N-cq^2) q_j + 2cn \leq N^2 + 2cn . \end{aligned}$$

To verify hypothesis (ii) it is sufficient to show that the estimate on the commutator holds for vectors in C^∞_{com}. This is also a simple computation:

$$\pm i[H,N] = \pm i[p^2,2cq^2] = \pm 4c(pq+qp) \leqslant 4c^{\frac{1}{2}}(p^2+cq^2) \leqslant 4c^{\frac{1}{2}}N .$$

In the proof of Theorem 12.1 the resolvent of the auxiliary operator played an important role. For the next theorem certain unitary groups are used in place of the resolvent.

Let $\mathcal{H} = L^2(\mathbb{R},dx)$. Let $p = -i\frac{d}{dx}$ and q = multiplication by x . Set

$W(s,t) = \exp(\frac{1}{2}ist)\exp(isq)\exp(itp) = \exp(-\frac{1}{2}ist)\exp(itp)\exp(isq)$ and

$W(u) = \iint u(s,t)W(s,t)dsdt$.

<u>Lemma</u> . If f is in $\mathcal{H}$ and u is in $\mathcal{S}(\mathbb{R}^2)$, then $W(u)f$ is in $\mathcal{S}(\mathbb{R}^1)$.

<u>Proof</u>: Let $\pi = i\frac{\partial}{\partial t} + \frac{s}{2}$ and $\chi = i\frac{\partial}{\partial s} - \frac{t}{2}$. Then $pW(u) = W(\pi u)$ and $qW(u) = W(\chi u)$. Hence $p^nW(u)f = W(\pi^n u)f$ and $q^nW(u)f = W(\chi^n u)f$ are in L^2 for all u , so $W(u)f$ is in $\mathcal{S}$.

<u>Theorem 12.2</u> . Let $\mathcal{H} = L^2(\mathbb{R},dx)$. Let A be an elliptic polynomial in the variables p and q . Assume that $A \geqslant 0$ as an operator acting on $\mathcal{S}$. Then A is essentially self-adjoint.

<u>Proof</u>: We may assume that $A \geqslant 1$. Consider g in $\mathcal{H}$ with $g \perp A\mathcal{S}$. We must show that $g = 0$.

We have $\langle g,AW(u)g\rangle = 0$ for all u in $\mathcal{S}$, by the lemma. But $\langle g,AW(u)g\rangle = \langle g,W(Lu)g\rangle = \iint \langle g,W(s,t)g\rangle(Lu)(s,t)dsdt$, where L is the same polynomial in π and χ . This says that $L\langle g,Wg\rangle = 0$ in the sense of distributions.

Now we appeal to the regularity theorem for elliptic partial differential equations. The conclusion is that $\langle g,W(s,t)g\rangle$ is a C^∞ function of s and t. In particular $p^n g$ and $q^n g$ are in L^2 for all n, so g is in $\mathcal{S}$.

But then it follows that $\langle g,Ag\rangle = 0$, so $g = 0$.

This result has an obvious extension to n dimensions.

NOTES

This type of commutator theorem originates with Glimm and Jaffe (1972b). The present version is due to Nelson (1972). The proof and the application follow a paper of Faris and Lavine (1974). (Kalf (1973) has shown that the application may also be treated by partial differential equation methods.)

The proof of Nelson (1972) actually gives a stronger result: If $N \geqslant 0$, $\mathcal{D}(H) \supset C^\omega(N)$ (the analytic vectors), and $\pm i[H,N] \leqslant cN$, then H is essentially self-adjoint. McBryan (1973) ( ) and Yakimov (1974) also have improvements on the theorem.

The result on elliptic polynomials in p and q is a special case of a theorem of Nelson and Stinespring (1959). The elliptic regularity theorem used in the proof may be found in the book of Dunford and Schwartz (1963). (The exact reference is to the last sentence of Corollary 4 on page 1708.)

Chernoff (1973) has given a rather different approach to self-adjointness questions based on finite propagation speed. Kato (1973) has used this to obtain a result for Schrödinger operators similar to Theorem 12.1 .

Part III : SELF-ADJOINT EXTENSIONS

§13 EXTENSIONS OF HERMITIAN OPERATORS

In this section we review the standard theory of extensions of Hermitian operators. The main conclusion is that if a Hermitian operator is not essentially self-adjoint, then either it has no self-adjoint extensions or it has infinitely many. In the latter case they are parametrized by unitary operators.

Let A be a densely defined operator with adjoint A^*. Since A^* is a closed operator, its domain $\mathcal{D}(A^*)$ is a Hilbert space with the inner product $\langle f,g\rangle_{\mathcal{D}} = \langle A^*f,A^*g\rangle + \langle f,g\rangle$.

Now let A be a Hermitian operator. A <u>boundary value</u> is a continuous linear functional on $\mathcal{D}(A^*)$ which vanishes on $\mathcal{D}(A)$. A <u>boundary condition</u> is a condition obtained by setting a boundary value equal to zero.

If A is a Hermitian operator, and A_1 is a self-adjoint extension of A, then A_1 is a restriction of A^*. Thus to specify A_1 it is enough to specify the subspace $\mathcal{D}(A_1) \subset \mathcal{D}(A^*)$. Since $\mathcal{D}(\bar{A}) \subset \mathcal{D}(A_1) \subset \mathcal{D}(A^*)$ as closed subspaces, it is possible to specify $\mathcal{D}(A_1)$ by imposing a set of boundary conditions.

Let A be a Hermitian operator and set $D_+ = \ker(A^*-i)$ and $D_- = \ker(A^*+i)$. Then A is essentially self-adjoint if and only if $D_+ = D_- = 0$. If A is closed, then A is self-adjoint if and only if $D_+ = D_- = 0$. The spaces D_+ and D_- thus measure the deviation from self-adjointness. They are called the <u>deficiency spaces</u>, and their dimensions are the <u>deficiency indices</u> of A.

If e is in D_+ and f is in D_-, then $A^*(e+f) = i(e-f)$. Hence for e_1 and e_2 in D_+ and f_1 and f_2 in D_- we have

$$\langle e_1+f_1, e_2+f_2\rangle_{\mathcal{D}} = 2(\langle e_1,e_2\rangle + \langle f_1,f_2\rangle) = \langle e_1,e_2\rangle_{\mathcal{D}} + \langle f_1,f_2\rangle_{\mathcal{D}} .$$

From this we see that $D_+ \perp D_-$ in the norm of $\mathcal{D}(A^*)$.

<u>Proposition 13.1</u>. Let A be a closed Hermitian operator. Then $\mathcal{D}(A^*) = \mathcal{D}(A) \oplus D_+ \oplus D_-$, where the direct sum decomposition is orthogonal with respect to the graph inner product of A^*.

<u>Proof</u>: Let g be orthogonal to $\mathcal{D}(A)$ in $\mathcal{D}(A^*)$. Then $\langle Af, A^*g\rangle + \langle f,g\rangle = 0$ for $f \in \mathcal{D}(A)$, so A^*g is in $\mathcal{D}(A^*)$ and $(A^{*2}+1)g = 0$. Thus $\mathcal{D}(A^*) = \mathcal{D}(A) \oplus \ker(A^{*2}+1)$, by the projection theorem.

Now consider g in $\ker(A^{*2}+1)$. We may write it as $g = \frac{1}{2i}((A^*+i)g - (A^*-i)g)$. Since $(A^*\pm i)g$ is in $D_\pm$, this shows that $\ker(A^{*2}+1) = D_+ \oplus D_-$.

Let A be a Hermitian operator. Define the sesquilinear form B on $\mathcal{D}(A^*)$ by $B(g,h) = \langle g, A^*h\rangle - \langle A^*g, h\rangle$. Then B is skew-Hermitian: $B(g,h) = -B(h,g)^*$. This form is called the <u>boundary form</u>.

Notice that if u_1 and u_2 are in $\mathcal{D}(A)$, e_1 and e_2 in D_+ and f_1 and f_2 in D_-, then

$B(u_1+e_1+f_1, u_2+e_2+f_2) = 2i(\langle e_1,e_2\rangle - \langle f_1,f_2\rangle)$.

<u>Corollary</u>. All boundary conditions are of the form $B(g,h) = 0$, where g is a fixed element of $D_+ \oplus D_-$.

<u>Proof</u>: It follows from the Riesz representation theorem that the most general boundary condition is of the form $\langle f,h\rangle_{\mathcal{D}} = 0$ for some fixed element f of $D_+ \oplus D_-$. But

$$\langle f,h\rangle_{\mathcal{D}} = \langle A^*f,A^*h\rangle + \langle f,h\rangle = \langle (A^{*2}+1)f,h\rangle + \langle A^*f,A^*h\rangle - \langle A^{*2}f,h\rangle = B(A^*f,h)$$

since $(A^{*2}+1)f = 0$. Thus we may set $g = A^*f$.

It is worth noting here that there are two possible norms on the deficiency spaces $D_\pm$, namely that of $\mathcal{D}(A^*)$ and that of the original Hilbert space. However they differ only by a constant factor.

<u>Proposition 13.2</u> . Let A be a Hermitian operator. Let A_1 be an operator with $A \subset A_1 \subset A^*$. Then A_1 is Hermitian if and only if $\mathcal{D}(A_1)$ is the direct sum of $\mathcal{D}(A)$ and the graph of an isometry from a linear subspace of D_+ to a linear subspace of D_- .

<u>Proof</u>: Consider e in D_+ and f in D_- such that $e + f$ is in $\mathcal{D}(A_1)$. Then A_1 is Hermitian when $B(e+f,e+f) = 0$. But since $A^*(e+f) = i(e-f)$, $B(e+f,e+f) = 2i(\|e\|^2-\|f\|^2)$.

<u>Theorem 13.3</u> . Let A be a closed Hermitian operator. Let A_1 be an operator with $A \subset A_1 \subset A^*$. Then A_1 is self-adjoint if and only if $\mathcal{D}(A_1)$ is the direct sum of $\mathcal{D}(A)$ and the graph of a unitary operator from D_+ to D_- .

<u>Proof</u>: Since A is closed, range$(A-i) \oplus D_+ = H$ and range$(A+i) \oplus D_- = H$. Let A_1 be a Hermitian extension of A . Then A_1 is self-adjoint if and only if range$(A_1-i) = H$ and range$(A_1+i) = H$, which amounts to the requirement that range$(A_1-i) \supset D_+$ and range$(A_1+i) \supset D_-$.

But if e and f are in D_+ and D_- with $e + f$ in $\mathcal{D}(A_1)$, then $(A_1-i)(e+f) = -2if$ and $(A_1+i)(e+f) = 2ig$, so this is true if and only if e and f can be arbitrary elements of D_+ and D_- . Thus the assertion follows from Proposition 13.2 .

Theorem 13.4 . Let A be a closed Hermitian operator. If A is not self-adjoint, then either A has no self-adjoint extensions or infinitely many self-adjoint extensions, according to whether the deficiency indices are unequal or equal.

Proof: Assume D_+ and D_- are not both zero. Then there are no unitary operators from D_+ to D_-, or infinitely many, according to whether the dimensions are unequal or equal.

Corollary . Let A be a Hermitian operator with a unique self-adjoint extension. Then A is essentially self-adjoint.

Proof: $\bar{A}$ also has a unique self-adjoint extension, and $\bar{A}$ is closed, so $\bar{A}$ is self-adjoint.

Lemma . Let A be a Hermitian operator. Let A_1 be an operator with $A \subset A_1 \subset A^*$. Then h is in $\mathcal{D}(A_1^*)$ if and only if h is in $\mathcal{D}(A^*)$ and $B(g,h) = 0$ for all g in $\mathcal{D}(A_1)$.

Proof: Since $A \subset A_1 \subset A^*$, it follows by taking adjoints that $A \subset A_1^* \subset A^*$. Thus if h is in $\mathcal{D}(A_1^*)$, then $A_1^* h = A^* h$.

It follows that h is in $\mathcal{D}(A_1^*)$ if and only if $\langle A_1 g,h\rangle = \langle g,A^* h\rangle$ for all g in $\mathcal{D}(A_1)$, that is, if and only if $B(g,h) = \langle g,A^* h\rangle - \langle A^* g,h\rangle = 0$ for all g in $\mathcal{D}(A_1)$.

Proposition 13.5 . Let A be a closed Hermitian operator. Let A_1 be a self-adjoint extension of A , so that $\mathcal{D}(A_1) = \mathcal{D}(A) \oplus U \subset \mathcal{D}(A^*)$, where U is the graph of a unitary operator from D_+ to D_- . Then the boundary conditions defining the set of h which belong to $\mathcal{D}(A_1)$ are given by $B(u,h) = 0$ for all u in U .

Proof: The set $\mathcal{D}(A_1)$ consists of all $g = f + u$ where f is in $\mathcal{D}(A_1)$ and u is in U. Thus by the lemma h is in $\mathcal{D}(A_1)$ if and only if $0 = B(g,h) = B(u,h)$ for all g in $\mathcal{D}(A_1)$, that is, for all u in U.

Proposition 13.6. Let W be a Hilbert space and $T : W \to W$ be a conjugation. Let A be a Hermitian operator which is real with respect to the conjugation T. Then A has a self-adjoint extension.

Proof: The conjugation T maps D_+ onto D_-, so the deficiency indices are equal.

EXAMPLE

Let $\mathcal{H} = L^2(\mathbb{R}^+, dx)$, where $\mathbb{R}^+ = [0,\infty)$. We will always consider $\frac{d}{dx}$ to be defined on absolutely continuous functions f in L^2 such that f' is also in L^2. Set $A = -\frac{d^2}{dx^2}$, defined on f with the additional restriction that $f(0) = 0$ and $f'(0) = 0$. Then A is a closed Hermitian operator.

The adjoint A^* is $-\frac{d^2}{dx^2}$ with no restrictions at zero. Notice that A^* is not Hermitian, in fact the boundary form is $B(f,g) = \langle f, A^*g\rangle - \langle A^*f, g\rangle = f(0)^*g'(0) - f'(0)^*g(0)$.

The kernel of $A^* - z$ is spanned by $\exp(-(-z)^{\frac{1}{2}}x)$. Thus the deficiency spaces $D_\pm$ (corresponding to $z = \pm i$) are spanned by $e_\pm$, where $e_\pm$ are exponential functions satisfying $e_\pm(0) = 1$ and $e_\pm'(0) = -\exp(\pm i\frac{\pi}{4})$.

It follows that $B(e_{\pm},h) = h'(0) + \exp(\pm i\frac{\pi}{4})h(0)$. Set $u = e_{+} + \exp(i\theta)e_{-}$. The boundary condition $B(u,h) = 0$ is thus equivalent to $\cos\frac{\theta}{2}\, h'(0) + \cos(\frac{\theta}{2}-\frac{\pi}{4})h(0) = 0$.

Thus we see that the most general boundary condition defining a self-adjoint extension is of the form $ah(0) + bh'(0) = 0$ with a and b real (and not both zero). From this point of view there is no reason to choose one extension over any other.

NOTES

The theory of extensions of Hermitian operators may be found in the book of Dunford and Schwartz (1963).

§14 THE FRIEDRICHS EXTENSION

In this section we see that a Hermitian operator which is bounded below always has self-adjoint extensions. Among these there is a maximal one, the Friedrichs extension.

Friedrichs Extension Theorem 14.1 . Let A be a positive operator acting in a Hilbert space $\mathcal{H}$. Then the form $\langle f, Ag \rangle$, defined for f and g in $\mathcal{D}(A)$, is closable.

Proof: Let $\mathcal{Q}$ be the completion of $\mathcal{D}(A)$ with respect to the norm $\|f\|_{\mathcal{D}}^2 = \langle f,(A+1)f \rangle$. The inclusion $\mathcal{D}(A) \subset \mathcal{H}$ extends by continuity to a map $\iota : \mathcal{Q} \to \mathcal{H}$. We must show that ι is injective, so that we may identify $\mathcal{Q} \subset \mathcal{H}$.

The point is that for fixed g in $\mathcal{D}(A)$ the functional $f \mapsto \langle f,g \rangle_{\mathcal{D}} = \langle f,(A+1)g \rangle$ is continuous in the norm of $\mathcal{H}$. Thus since $\iota : \mathcal{Q} \to \mathcal{H}$ is continuous, we have $\langle f,g \rangle_{\mathcal{D}} = \langle \iota f,(A+1)g \rangle$ for al f in $\mathcal{Q}$. Hence if $\iota f = 0$, then $\langle f,g \rangle_{\mathcal{D}} = 0$ for all g in $\mathcal{D}$. Since $\mathcal{D}$ is dense in $\mathcal{Q}$, this implies $f = 0$. Hence $\iota : \mathcal{Q} \to \mathcal{H}$ is injective.

Thus any positive Hermitian operator A acting in $\mathcal{H}$ determines a closed densely defined positive form. On the other hand, for any such form the associated operator A_F is self-adjoint. This self-adjoint operator A_F is called the Friedrichs extension of A .

The Friedrichs extension of a positive Hermitian operator is thus a positive self-adjoint operator. One can define the Friedrichs extension of an arbitrary semibounded Hermitian operator. If $A \geq d$ is such an operator, then its Friedrichs extension is a self-adjoint operator $A_F \geq d$, with the same lower bound. Notice that if A is already self-adjoint, then the Friedrichs extension of A is A itself.

The Friedrichs extension is maximal among semibounded self-adjoint extensions.

<u>Proposition 14.2</u> . Let A be a positive densely defined operator and let A_F be its Friedrichs extension. Let A_1 be any other self-adjoint extension which is bounded below. Then $A_1 \leq A_F$. In fact, the form of A_1 extends the form of A_F .

<u>Proof</u>: Let $-c$ be strictly less than the lower bound of A_1 . By the construction of the Friedrichs extension, $\mathcal{Q}(A_F)$ is the completion of $\mathcal{D}(A)$ in the norm $\langle f,(A+1)f\rangle$, or what is the same, the completion in the norm $\langle f,(A+c)f\rangle = \langle f,(A_1+c)f\rangle$. On the other hand $\mathcal{Q}(A_1)$ is the completion of the larger space $\mathcal{D}(A_1) \supset \mathcal{D}(A)$ in the same norm $\langle f,(A_1+c)f\rangle$. Hence $\mathcal{Q}(A_F) \subset \mathcal{Q}(A_1)$ and $\langle f,A_1 f\rangle = \langle f,A_F f\rangle$ for all f in $\mathcal{Q}(A_F)$.

<u>EXAMPLE</u>

Let $\mathcal{H} = L^2(\mathbb{R}^+,dx)$ and A be $-\frac{d^2}{dx^2}$ with boundary conditions $f(0) = 0$ and $f'(0) = 0$. Then $A \geq 0$, so A has a Friedrichs extension A_F .

Notice that $\langle f,Af\rangle = \langle \frac{d}{dx}f,\frac{d}{dx}f\rangle$ for all f in $\mathcal{D}(A)$. The completion $\mathcal{Q}$ of $\mathcal{D}(A)$ thus consists of the f in $\mathcal{H}$ such that f' is in $\mathcal{H}$ and such that $f(0) = 0$. (The condition $f'(0) = 0$ is not preserved in the completion.)

To find the self-adjoint operator A_F , we ask for which f in $\mathcal{Q}$ is $g \mapsto \langle \frac{d}{dx}g, \frac{d}{dx}f \rangle$ continuous on $\mathcal{H}$. Since every f in $\mathcal{Q}$ satisfies $f(0) = 0$, the only additional restriction is that f'' is in H . Thus $A_F = -\frac{d^2}{dx^2}$ with the boundary condition $f(0) = 0$.

It is interesting to see what self-adjoint operator is associated with the form $\langle \frac{d}{dx}f, \frac{d}{dx}f \rangle$ defined on all f in H such that f' is also in H . We ask for which f among these is $g \mapsto \langle \frac{d}{dx}g, \frac{d}{dx}f \rangle$ continuous on H . Since $\langle g', f' \rangle = -\langle g, f'' \rangle - g(0)^* f'(0)$, we see that it is necessary that $f'(0) = 0$. Hence the operator is $-\frac{d^2}{dx^2}$ with the boundary condition $f'(0) = 0$.

NOTES

The Friedrichs extension is treated briefly in Nelson (1969) and more extensively in Kato (1966).

§15 EXTENSIONS OF SEMI-BOUNDED OPERATORS

Let A be a semi-bounded Hermitian operator. The Friedrichs extension A_F of A is a semi-bounded self-adjoint operator with the same lower bound. The purpose of this section is to classify all self-adjoint extensions A_1 of A which are bounded below by some fixed constant. We will find that they are parametrized by closed positive forms. This has the advantage that the closed positive forms are a partially ordered set.

We consider for simplicity the case $A \geq d > 0$ and classify all self-adjoint extensions $A_1 \geq 0$. It is useful to define the <u>deficiency space</u> of A as $N = \ker A^*$.

Let $\mathcal{H}$ be a Hilbert space and let S be a closed positive form on some linear subspace of $\mathcal{H}$. Let $\mathcal{K}$ be the closure of the domain of S. Then by the representation theorem S is the form of a positive self-adjoint operator acting in $\mathcal{K}$. Hence by the spectral theorem S is isomorphic to the form defined by a function on some measure space with values in $[0,\infty]$. (The set on which the function is $+\infty$ corresponds to $\mathcal{K}^\perp$.)

It follows from this that if S is a closed positive form, then so is S^{-1}. (In fact both are isomorphic to the form defined by a function with values in $[0,\infty]$.) Note that $SS^{-1} = 1$ wherever neither S nor S^{-1} are zero.

<u>Theorem 15.1</u>. Let A be a Hermitian operator such that $A \geq d > 0$. Let $N = \ker A^*$. There is a bijective correspondence between positive self-adjoint extensions A_1 of A and positive closed forms G defined in N, given by

$$A_1^{-1} = A_F^{-1} + G^{-1} .$$

Proof: Let A_1 be a positive self-adjoint extension of A. Then A_1 defines a positive closed form, so A_1^{-1} is also a positive closed form. Since $A_1 \leq A_F$, $A_F^{-1} \leq A_1^{-1}$. But $A_F \geq d > 0$, so $A_F^{-1} \leq d^{-1}$ is bounded. Hence $G^{-1} = A_1^{-1} - A_F^{-1}$ is also a positive closed form. Since $N^{\perp} = \text{range}\,\bar{A} \subset \text{range}\,A_1$, $A_1^{-1}A_1 = 1$ on $N^{\perp}$. Thus since A_1 and A_F are both extensions of $\bar{A}$, $G^{-1} = 0$ on $N^{\perp}$.

On the other hand, let G be a positive closed form in N. Then G^{-1} is a positive closed form which is zero on $N^{\perp}$. Since A_F^{-1} is positive and bounded, $A_1^{-1} = A_F^{-1} + G^{-1}$ is also a a positive closed form. Since A_F^{-1} is never zero, neither is A_1^{-1}, so A_1 is a densely defined positive closed form, that is, a self-adjoint operator. Since G^{-1} is zero on $N^{\perp}$, $A_1\bar{A}^{-1} = A_1A_F^{-1} = A_1A_1^{-1}$ on $N^{\perp}$, and since $N^{\perp} \subset \text{range}\,A_1$, this is 1 on $N^{\perp}$. Hence A_1 extends $\bar{A}$.

Corollary. Let A be a closed Hermitian operator with $A \geq d > 0$. Assume that A is not self-adjoint. Then there is a self-adjoint extension A_1 of A which is minimal among all positive self-adjoint extensions, and zero is an eigenvalue of A_1.

Proof: Take $G = 0$ on N. Then $G^{-1} = \infty$ on N, so the extension A_1 satisfies $A_1^{-1} = A_F^{-1}$ on $N^{\perp}$ and $A_1 = 0$ on N. Since A is not self-adjoint, $N \neq 0$ and so zero is an eigenvalue.

Notice that the Friedrichs extension corresponds to the choice $G = \infty$.

If A is a closed Hermitian operator with $A \geq m$ which is not self-adjoint, then for every $c < m$, $N_c = \ker(A^*-c) \neq 0$. Thus we may apply the above construction to $A - c$ to produce a self-adjoint extension of A which is bounded below by c and which has c as an eigenvalue. It follows in particular that if a semi-bounded Hermitian operator has a unique semi-bounded extension, then the operator is essentially self-adjoint.

Proposition 15.2 . Let A be a Hermitian operator such that $A \geq d > 0$. Let $N = \ker A^*$. Then $\mathcal{Q}(A_F) \cap N = 0$.

Proof: Let u be in N . Then $\langle u,Av\rangle = 0$ for all v in $\mathcal{D}(A)$. If u is also in $\mathcal{Q}(A_F)$, this says that $\langle u,A_Fv\rangle = 0$ for all v in $\mathcal{D}(A)$. But $\mathcal{D}(A)$ is dense in $\mathcal{Q}(A_F)$, so this implies that $u = 0$.

Theorem 15.3 . Let A be a Hermitian operator such that $A \geq d > 0$. Let $N = \ker A^*$. Let A_1 be a positive self-adjoint extension of A and let G be the corresponding closed positive form in N . Then A_1 is the direct sum of A_F and G , in the sense that $\langle g,A_1g\rangle = \langle f,A_Ff\rangle + \langle u,Gu\rangle$, where g is in $\mathcal{Q}(A_1)$, $g = f + u$ with f in $\mathcal{Q}(A_F)$ and u in $\mathcal{Q}(G)$.

Proof: We have $\langle h,A_1^{-1}h\rangle = \langle h,A_F^{-1}h\rangle + \langle h,G^{-1}h\rangle$ for all h in $\mathcal{Q}(A_1^{-1})$. This continues to hold for all h in the completion with respect to the norm $\langle h,A_1^{-1}h\rangle$. Assume now that h is orthogonal to $\ker A_1 = \ker G$ and set $g = A_1^{-1}h$, $f = A_F^{-1}h$, and $u = G^{-1}h$. Then $g = f + u$ and the form equality becomes $\langle g,A_1g\rangle = \langle f,A_Ff\rangle + \langle u,Gu\rangle$. Notice that g and u are restricted to be orthogonal to $\ker A_1$.

If w is in $\ker A_1$, then $g + w = f + (u+w)$, and so the representation in fact holds in general.

EXAMPLE

Let $\mathcal{H} = L^2(\mathbb{R}^+, dx)$. Let $k > 0$ and $A = -\frac{d^2}{dx^2} + k^2$ with the boundary conditions $g(0) = 0$, $g'(0) = 0$. We wish to find all positive self-adjoint extensions.

Let $N = \ker A^*$. Since $A^* = -\frac{d^2}{dx^2} + k^2$ with no boundary conditions, N is spanned by $c \exp(-kx)$. Thus N is one-dimensional.

Let A_1 be a positive self-adjoint extension. The form of A_1 has the representation

$$\langle g, A_1 g\rangle = \langle f, A_F f\rangle + \langle u, Gu\rangle ,$$

where $g = f + u$ with $f(0) = 0$ and $u(x) = c \exp(-kx)$. Here $\langle f, A_F f\rangle = \langle f', f'\rangle + k^2\langle f, f\rangle$ and $\langle u, Gu\rangle = \ell|c|^2$ for some constant ℓ with $0 \leqslant \ell \leqslant +\infty$.

The decomposition $g = f + u$ may be made explicit by writing $g(x) = g(x) - g(0)\exp(-kx) + g(0)\exp(-kx)$. Thus $u(x) = g(0)\exp(-kx)$.

It is not difficult to compute that the form of A_1 is $\langle g, A_1 g\rangle = \langle g', g'\rangle + k^2\langle g, g\rangle + (\ell-k)|g(0)|^2$. The operator A_1 is defined on all g such that $\langle h, A_1 g\rangle = -\langle h, g''\rangle + k^2\langle h, g\rangle + h(0)^*((\ell-k)g(0) - g'(0))$ is continuous in h . Thus A_1 is $-\frac{d^2}{dx^2} + k^2$ acting on g with the boundary condition $g'(0) = (\ell-k)g(0)$.

The case $\ell = \infty$ gives the Friedrichs extension $A_F = -\frac{d^2}{dx^2} + k^2$ with the boundary condition $g(0) = 0$. If $\ell \geqslant k$ then $A_1 \geqslant k^2$, while if $\ell < k$, then A_1 has an eigenvalue strictly less than k^2 .

(This shows, incidentally, that the Friedrichs extension A_F is not characterized by the fact that it has the same lower bound as A .)

EXAMPLE

Let $H = L^2(\mathbb{R}^+, dx)$. Let $A_o = -\frac{d^2}{dx^2} + \frac{a}{x^2}$, defined on functions which vanish near zero. Let $A = \bar{A}_o$. Then $A^* = -\frac{d^2}{dx^2} + \frac{a}{x^2}$ with no boundary conditions.

The behavior of these operators depends on the value of the parameter a . We distinguish three cases:

1) $\frac{3}{4} \leq a$

2) $-\frac{1}{4} \leq a < \frac{3}{4}$

3) $a < -\frac{1}{4}$

Cases 1 and 2 are distinguished by the fact that A is positive. This may be seen by noting that if f vanishes near zero, then $0 \leq \langle(\frac{d}{dx}-\frac{1}{2x})f, (\frac{d}{dx}-\frac{1}{2x})f\rangle = \langle f, (-\frac{d^2}{dx^2}-\frac{1}{4x^2})f\rangle$. Thus $A_o \geq 0$, so $A \geq 0$. It follows that the Friedrichs extension $A_F \geq 0$ exists. Thus A has a canonical extension which is a positive self-adjoint operator.

Case 1 is distinguished by the fact that A is already self-adjoint; there is no choice at all of self-adjoint extension. This may be seen by computing $N = \ker(A^*+1)$. The equation $-f'' + \frac{a}{x^2}f + f = 0$ may be solved in terms of Hankel functions, but there is a more elementary way to see what is happening. When $x \sim \infty$ the solutions behave like $c_1\exp(x) + c_2\exp(-x)$. The only solutions in L^2 at infinity are of the form $f(x) \sim c\exp(-x)$. On the other hand, such a solution behaves like $f(x) \sim d_1x^{r_1} + d_2x^{r_2}$ when $x \sim 0$.

Here r_1 and r_2 are the solutions of $r(r-1) = a$. If $a \geq \frac{3}{4}$, then $r_1 \geq \frac{3}{2}$ and $r_2 \leq -\frac{1}{2}$. Since x^{r_2} is not in L^2 near zero, f is not in L^2 and so $N = 0$.

In case 2 , $\frac{3}{2} > r_1 > \frac{1}{2} > r_2 > -\frac{1}{2}$ (except when $a = -\frac{1}{4}$, $r = r_1 = \frac{1}{2}$ and the solution $f(x) \sim d_1 x^{\frac{1}{2}} + d_2 x^{\frac{1}{2}} \log x$ near zero). Hence $f(x)$ is in L^2 , N is one dimensional, and there are many self-adjoint extensions. The Friedrichs extension is characterized by the fact that functions in its domain vanish at $x = 0$ like x^{r_1}.

A similar analysis in case 3 shows that the solutions of $(A^*-i)f = 0$ behave at the origin like $d_1 x^{r_1} + d_2 x^{r_2}$ where r_1 and r_2 both satisfy $\operatorname{Re} r = \frac{1}{2}$. Hence there is a one dimensional space of solutions and A is not self-adjoint. But what is worse, there is no natural way to select a self-adjoint extension. The self-adjoint extensions are characterized by a phase $\exp(i\theta)$, and this is an additional piece of information needed to specify the operator.

EXAMPLE

It is worth looking at an example in n dimensions. Let $\mathcal{H} = L^2(\mathbb{R}^n, dx)$ and consider the operator $A_o = -\Delta - \frac{(n-2)^2}{4} \frac{1}{r^2}$ acting on functions which vanish near the origin. Here $r = |x|$ is the radial distance from the origin. Since $0 \leq \| (\frac{\partial}{\partial r} + \frac{n-2}{r}) f \|^2 = \langle \frac{\partial}{\partial r} f, \frac{\partial}{\partial r} f \rangle - \frac{(n-2)^2}{4} \langle f, \frac{1}{r^2} f \rangle \leq \langle f, A_o f \rangle$, A_o is positive. Hence it has a Friedrichs extension A_F which is also positive.

If $n \geq 2$, then the functions which vanish near the origin are dense in $\mathcal{Q}(\Delta)$, so $A_F \leq -\Delta$. However notice the curious fact that when $n = 1$, it is false that $A_F \leqslant -\frac{d^2}{dx^2}$. In fact, when $n = 1$ functions in $\mathcal{Q}(A_F)$ vanish at the origin, which is not necessarily so for functions in $\mathcal{Q}(-\frac{d^2}{dx^2})$.

The same reasoning shows that if $n \geqslant 2$ and if W is a real function on $\mathbb{R}^n$ with $W(x) \geqslant -\frac{(n-2)^2}{4}\frac{1}{r^2}$, then

$$0 \leqslant A_F \leqslant -\Delta + W .$$

In other words, when $n \geqslant 3$ a shallow attractive potential doesn't produce negative eigenvalues. (However when $n = 1$ or 2 the argument fails; an attractive potential always produces a negative eigenvalue.)

NOTES

The theory of extensions of semi-bounded operators is classic work of Krein (1947). The theory was extended by Birman (1956).

Nelson (1964) found a canonical choice of boundary condition for $-\frac{d^2}{dx^2} + \frac{a}{x^2}$, $a < -\frac{1}{4}$, that makes it a non-self-adjoint operator. The relationship between Nelson's extension and the self-adjoint extensions is discussed by Radin ( ).

There is a rigorous general theory of the relation between boundary conditions and self-adjoint extensions of ordinary differential operators. This Weyl theory justifies the procedure of examining the behavior of the solutions at the two end points separately. There is a good discussion in the book of Dunford and Schwartz (1963).

Part IV : MOMENTS

§16 ANALYTIC VECTORS

If A is a self-adjoint operator and f is a unit vector, the moments of A are the numbers $\langle f, A^n f\rangle$, $n = 1,2,3,\ldots$. One might hope that if one knows the moments of A for sufficiently many f , this would determine A . This procedure might also give a way of showing that a Hermitian operator is essentially self-adjoint. In fact this is so, but it is necessary to make precise what is meant by sufficiently many f .

Let A be an operator. The space of C^∞ vectors for A is the space $C^\infty(A)$ consisting of all f in $\mathcal{H}$ such that $A^n f$ is in $\mathcal{D}(A)$ for all $n = 0,1,2,3,\ldots$. If f is a C^∞ vector for A , then the moments $\langle f, A^n f\rangle$ are all defined.

The space $C^\omega(A)$ of analytic vectors is the subspace of $C^\infty(A)$ consisting of all f such that there exists constants a and b with $\|A^n f\| \leq ab^n n!$.

The space $\mathcal{B}(A)$ of bounded vectors is the subspace of $C^\infty(A)$ consisting of all vectors which satisfy an estimate $\|A^n f\| \leq ab^n$. Clearly $\mathcal{B}(A) \subset C^\omega(A) \subset C^\infty(A)$.

If A is a self-adjoint operator, it is clear from the spectral theorem that $\mathcal{B}(A)$ is dense in $\mathcal{H}$. (The converse is also true: if A is a closed Hermitian operator and $\mathcal{B}(A)$ is dense in $\mathcal{H}$, then A is self-adjoint. In fact, if h is in $\mathcal{B}(A)$ and $|z| > b$, then $h = (z-A)g$, where $g = (z-A)^{-1}h = z\Sigma_n (A/z)^n h$. Thus $\mathrm{range}(z-A) \supset \mathcal{B}(A)$ is dense in $\mathcal{H}$, and since A is closed, $\mathrm{range}(z-A) = \mathcal{H}$. However we will see later that there is a better result: if A is a closed Hermitian operator and $C^{\omega}(A)$ is dense in H, then A is self-adjoint.)

Let A be a Hermitian operator acting in H and consider f in $C^{\infty}(A)$. Let $D(f)$ be the linear span of the $A^k f$, $k = 0,1,2,\ldots$ Let $M(f)$ be the closure of $D(f)$ in H. Then the restriction $A : D(f) \to M(f)$ is a Hermitian operator acting in $M(f)$.

This restricted operator has a self-adjoint extension. In fact, let $T : D(f) \to D(f)$ be defined by $T\Sigma c_k A^k f = \Sigma c_k^* A^k f$, and $T : M(f) \to M(f)$ be the extension by continuity. Then T is a conjugation which commutes with A.

The vector f is said to be a <u>determining vector</u> if $A : D(f) \to M(f)$ is essentially self-adjoint. (This is equivalent to the uniqueness of the self-adjoint extension.)

<u>Proposition 16.1</u>. Let A be a Hermitian operator acting in H. Assume that the union of the $D(f)$, where f is a determining vector, is dense in H. Then A is essentially self-adjoint.

<u>Proof</u>: Let h be in H. Then there is a determining vector f and a u in $D(f)$ such that u approximates h. But since f determines, there is a g in $D(f)$ such that $(A\pm i)g$ approximates u. Hence $(A\pm i)g$ approximates h. This shows that range $A\pm i$ dense in H.

<u>Proposition 16.2</u>. Let A be a Hermitian operator. Then every analytic vector for A is a determining vector for A.

<u>Proof</u>: Let f be in $C^{\omega}(A)$. The Hermitian operator $A : D(f) \to M(f)$ has a self-adjoint extension A_1 acting in $M(f)$.

Since f is in $C^{\omega}(A)$, there is an $\varepsilon > 0$ such that the expansion of $\exp(itA_1)f = \exp(itA)f$ converges for $|t| < \varepsilon$.

Let g be in $D(f)$. Then $\exp(itA_1)g = \exp(itA)g$ also converges for $|t| < \varepsilon$. Since $D(f)$ is dense in $M(f)$, $\exp(itA_1)$ is uniquely determined by A . Hence A_1 is uniquely determined by A , so A acting in $M(f)$ is essentially self-adjoint on $D(f)$.

Theorem 16.3 . Let A be a Hermitian operator. If A has a dense set of analytic vectors, then A is essentially self-adjoint.

Proof: If A has a dense set of analytic vectors, then A has a dense set of determining vectors, and so A is essentially self-adjoint.

The rest of this section is devoted to the relation between the analytic vectors of two different operators. We say that A analytically dominates X if $C^{\omega}(A) \subset C^{\omega}(X)$. In order to get analytic dominance, we will need estimates on commutators. If X and A are operators, we write $(\mathrm{ad}X)A = XA - AX$ for the operation of taking the commutator of A with X . The following theorem requires second order estimates on commutators.

Theorem 16.4 . Let A and X be operators. Assume that there are constants a , b and c such that for all u in $\mathcal{D}(A)$

$$\|Xu\| \leq c\|Au\|$$

and

$$\|(\mathrm{ad}X)^n Au\| \leq ab^n n!\|Au\| .$$

Then A analytically dominates X .

Proof: We first bound $\|X^n u\|$ in terms of a linear combination $\pi_n(u)$ of $\|Au\|, \|A^2u\|, \ldots, \|A^n u\|$. In order to apply the commutator bounds, it is convenient to first prove that $c\|AX^{n-1}u\| \leq \pi_n(u)$. It will follow from this that $\|X^n u\| \leq \pi_n(u)$.

Since $X^n Au = \sum_{k=0}^{n} \binom{n}{k} ((\mathrm{ad}X)^k A) X^{n-k} u$, we may write

$$AX^n u = X^n Au - \sum_{k=1}^{n} \binom{n}{k} ((\mathrm{ad}X)^k A) X^{n-k} u$$

and try to bound the sum. This gives

$$c\|AX^n u\| \leq c^2 \|AX^{n-1}Au\| + c\sum_{k=1}^{n} \binom{n}{k} ab^k k! \|AX^{n-k}u\| .$$

Define $\pi_n(u)$ inductively by $\pi_0(u) = \|u\|$ and $\pi_{n+1}(u) = c\pi_n(Au) + \sum_{k=1}^{n} \binom{n}{k} ab^k k! \pi_{n-k+1}(k)$. It follows that if $c\|AX^j u\| \leq \pi_{j+1}(u)$ for $j = 0,2,\ldots,n-1$, then $c\|AX^n\| \leq \pi_{n+1}(u)$. Hence $\pi_n(u)$ is the desired bound. The only difficulty is that it is defined in terms of a complicated recursion relation.

The recursion relation for $\frac{1}{n!}\pi_n(u)$ involves a convolution, so it is convenient to introduce the transform $\phi(t,u) = \sum_n \frac{t^n}{n!}\pi_n(u)$. The relation then becomes

$$\frac{d}{dt}\phi(t,u) = c\phi(t,Au) + \nu(t)\frac{d}{dt}\phi(t,u) ,$$

where $\nu(t) = abt(1-bt)^{-1}$. Thus the convolution is replaced by multiplication by $\nu(t)$ and the shift by differentiation.

Write $\phi(t,u) = \sum_r p_r(t)\|A^r u\|$. Then $\frac{d}{dt}p_r(t) = c\, p_{r-1}(t) + \nu(t)\frac{d}{dt}p_r(t)$. Since $p_r(0) = 1$ when $r \geq 1$, it follows that $p_r(s) = c\int_0^s (1-\nu(t))^{-1} p_{r-1}(t)dt$. Since $p_o(t) = 1$ this may be solved to give $p_r(s) = \frac{c^r \kappa(s)^r}{r!}$, where $\kappa(s) = \int_0^s (1-\nu(t))^{-1}dt$. Hence $\phi(t,u) = \sum \frac{c^r\kappa(s)^r}{r!}\|A^r u\|$.

If we put together the information we have up to now, we obtain the estimate

$$\sum \frac{s^n}{n!}\|X^n u\| \leq \sum_n \frac{c^n \kappa(s)^n}{n!}\|A^n u\| .$$

Since $\nu(0) = 0$, $\kappa(s) \to 0$ as $s \to 0$. Hence if u is in $C^\omega(A)$, the right hand side converges for sufficiently small s , and hence so does the left hand side. Thus u is in $C^\omega(X)$.

One situation which may occur is that A is a self-adjoint operator and X leaves $C^\infty(A)$ invariant. Since $C^\omega(A) \subset C^\infty(A)$, it is sufficient to obtain the estimates for X and A restricted to $C^\infty(A)$ in order to conclude that $C^\omega(A) \subset C^\omega(X)$. We take such a situation as the setting for the next theorem. This theorem says that first order estimates on commutators imply analytic dominance.

Theorem 16.5 . Let $H \geq 1$ be a self-adjoint operator. Let X be a Hermitian operator which leaves $C^\infty(H)$ invariant. Assume that $X \leq cH$ and $(\mathrm{ad}X)^n H \leq ab^n n! H$ as quadratic forms on $C^\infty(H)$. Then H analytically dominates X .

Proof: Let $\mathcal{Q} \subset \mathcal{H} \subset \mathcal{Q}^*$ be defined by $\|u\|_{\mathcal{Q}} = \|H^{\frac{1}{2}}u\|$ and $\|v\|_{\mathcal{Q}^*} = \|H^{-\frac{1}{2}}v\|$. Then $H : \mathcal{Q} \to \mathcal{Q}^*$ is an isomorphism. Since X is Hermitian and $(\mathrm{ad}X)^n H$ is either Hermitian or skew-Hermitian, it follows from the hypotheses that $X : \mathcal{Q} \to \mathcal{Q}^*$ and $(\mathrm{ad}X)^k H : \mathcal{Q} \to \mathcal{Q}^*$ are operators bounded by c and $ab^k k!$ respectively.

Let u be in $C^\omega(H)$. We wish to estimate $\|X^n u\|$. Set $w = Hu$ and write

$$\|X^n u\| = \|X^n H^{-1} w\| \leq \|H^{-1} X^n w\| + \|[X^n, H^{-1}] w\| \leq \|X^n w\|_{\mathcal{Q}^*} + \|[X^n, H^{-1}] w\|_{\mathcal{Q}} .$$

The point is that it is much easier to estimate $\|X^n w\|_{\mathcal{Q}^*}$. Simply apply the previous theorem to the operators X and H acting in $\mathcal{Q}^*$. The conclusion is that w is an analytic vector for X acting in $\mathcal{Q}^*$, that is, there is an estimate $\|X^m w\|_{\mathcal{Q}^*} \leq r\, s^m m!$.

The remaining term is more difficult. Let $\gamma > 0$ be a small constant and choose q so that $b \leq \gamma q$ and $s \leq q$. We will see that $\|[X^n, H^{-1}]w\|_{\mathcal{Q}} \leq q^n n!$.

Write $[X^n, H^{-1}] = -H^{-1}[X^n, H]H^{-1}$ and $[X^n, H] = \sum_{k=1}^{n} \binom{n}{k}((\mathrm{ad}X)^k H)X^{n-k}$. Insert the second identity in the first. The resulting factor on the right may be expressed as $X^{n-k}H^{-1} = H^{-1}X^{n-k} + [X^{n-k}, H^{-1}]$. This gives two terms to be estimated.

The estimate is done by induction. Assume that $\|[X^m, H^{-1}]w\|_{\mathcal{Q}} \leq q^m m!$ for $m < n$. Then

$$\|[X^n, H^{-1}]w\|_{\mathcal{Q}} \leq \sum_{k=1}^{n} \binom{n}{k} a b^k k! (r s^{n-k}(n-k)! + q^{n-k}(n-k)!)$$

$$\leq q^n n! \gamma (1-\gamma)^{-1} a(r+1) \leq q^n n!$$

if γ is chosen sufficiently small.

These estimates combine to give a bound for $\|X^n u\|$ which proves that u is in $C^{\omega}(X)$.

NOTES

The theorem on analytic vectors and self-adjointness is due to Nelson (1959). The proof given here follows Simon (1972); it is a modification of techniques of Nussbaum (1965).

There is a notion of quasi-analytic vector which generalizes the notion of analytic vector. Nussbaum (1965) has proved a stronger theorem relating quasi-analytic vectors to self-adjointness.

The original theorem on analytic dominance is also due to Nelson (1959). Goodman (1969) has shown that the first estimate in the hypothesis may be weakened. The other theorem on analytic dominance, which allows form estimates in the hypotheses, is due to Sloan ( ). Sloan ( ) has shown that his first estimate may also be weakened.

§17 SEMI-ANALYTIC VECTORS

Let A be a Hermitian operator. The space of semi-analytic vectors consists of the f in $C^\infty(A)$ which satisfy an estimate $\|A^n f\| \leq ab^n (2n)!$ for some constants a and b (depending on f). Every analytic vector is semi-analytic.

Lemma . Let f be a semi-analytic vector for A and let c be a constant. Then f is semi-analytic for $A + c$.

Proof:

$$\|(A+c)^n f\| \leq \sum_k \binom{n}{k} |c|^{n-k} \|A^k f\| \leq \sum_k \binom{n}{k} |c|^{n-k} ab^k (2k)! = a(b+|c|)^k (2k)!.$$

Proposition 17.1 . Let A be a semi-bounded Hermitian operator. Then every semi-analytic vector for A is a determining vector for A .

Proof: Let f be a semi-analytic vector for A . Let $D(f)$ be the linear span of the $A^k f$ and let $M(f)$ be the closure of $D(f)$. Let A_o be the restriction of A to $D(f)$. We must show that A_o acting in $M(f)$ is essentially self-adjoint.

Since A_o is semi-bounded, it has a self-adjoint extension. By the lemma we may assume that $A_o \geq m > 0$. According to the theory of extensions of semi-bounded operators it is enough to show that there is a unique self-adjoint extension $A_1 \geq 0$.

Since f is a semi-analytic vector, if t is sufficiently small the series

$$\cos(t\, A_1^{\frac{1}{2}})g = \sum_n (-1)^n (t^{2n}/(2n)!) A_o^n g$$

converges for all g in $D(f)$. Thus $\cos(t\, A_1^{\frac{1}{2}})$ is uniquely determined by A_o .

The value of $\frac{d^2}{dt^2} \cos(t\, A_1^{\frac{1}{2}})h$ at $t = 0$ is $-A_1 h$. Thus A_1 is also uniquely determined by A_o .

Theorem 17.2 . Let A be a semi-bounded Hermitian operator. Assume there exists a dense set of semi-analytic vectors for A . Then A is essentially self-adjoint.

EXAMPLE

The restriction of a self-adjoint operator to a dense set of analytic vectors need not be essentially self-adjoint. The difficulty is that the vectors need not be analytic for the restriction.

Let A be an unbounded self-adjoint operator with domain $\mathcal{D}$. Let $\mathcal{N} \subset \mathcal{D}$ be a closed hyperplane which is dense in $\mathcal{H}$. Let $\mathcal{E} = \mathcal{B}(A) \cap \mathcal{N}$. By the lemma below, $\mathcal{E}$ is dense in $\mathcal{N}$, and hence in $\mathcal{H}$. On the other hand, $\mathcal{E}$ is not dense in $\mathcal{D}$. So $\mathcal{E}$ is a dense set of analytic vectors for A , but the restriction of A to $\mathcal{E}$ is not essentially self-adjoint.

Lemma . Let $\mathcal{D}$ be a Hilbert space and $\mathcal{N}$ be a closed hyperplane. Let $\mathcal{B}$ be a dense linear subspace of $\mathcal{D}$. Then $\mathcal{B} \cap \mathcal{N}$ is dense in $\mathcal{N}$.

NOTES

The proof of the theorem on semi-analytic vectors is adapted from Simon (1971b). There is a more general notion, that of Stieltjes vector, and Nussbaum (1969) proved a stronger theorem relating Stieltjes vectors to self-adjointness. Chernoff (1972a) has discussed the relation between various self-adjointness results of this type. The idea for the example is due to Simon (unpublished). The density lemma may be found in the book Dynamical Theories of Brownian Motion, by E. Nelson (Princeton University Press, Princeton, N.J., 1967).

Masson and McClary (1972) have applied these notions to quantum mechanical Hamiltonians.

REFERENCES

Birman, M.Sh. (1956), On the theory of self-adjoint extensions of positive definite operators, Math. Sbornik 38, 431-450 (Russian).

Chernoff, P.R. (1970), Semigroup product formulas and addition of unbounded operators, Bull. Amer. Math. Soc. 76, 395-398.

Chernoff, P.R. (1972a), Perturbation of dissipative operators with relative bound one, Proc. Amer. Math. Soc. 33, 72-74.

Chernoff, P.R. (1972b), Some remarks on quasi-analytic vectors, Trans. Amer. Math. Soc. 167, 105-113.

Chernoff, P.R. (1973), Essential self-adjointness of powers of generators of hyperbolic equations, J. Functional Anal. 12, 401-414.

Chernoff, P.R. (1974), Product Formulas, Nonlinear Semigroups and Addition of Unbounded Operators, Mem. Amer. Math. Soc. #140, Amer. Math. Soc., Providence, R.I..

Davies, E.B. (1973), Properties of the Green's functions of some Schrödinger operators, J. London Math. Soc. (2), 7, 483-491.

Deny, J. (1950), Les potentiels d'énergie finie, Acta Math. 82, 107-183.

Devinatz, A. (1973), The deficiency index problem for ordinary self-adjoint differential operators, Bull. Amer. Math. Soc. 79, 1109-1127.

Dunford, N. and J.T. Schwartz (1963), Linear Operators, Part II, Interscience, New York.

Everitt, W.N. and M. Giertz (1972), Some inequalities associated with certain ordinary differential operators, Math. Z. 126, 308-326.

Everitt, W.N. and M. Giertz (1973), Inequalities and separation for certain partial differential expressions, preprint, Dept. of Math., Royal Inst. of Technology, Stockholm.

Faris, W. (1972a), Invariant cones and uniqueness of the ground state for fermion systems, J. Mathematical Phys. 13, 1285-1290.

Faris, W. (1972b), Quadratic forms and essential self-adjointness, Helv. Phys. Acta 45, 1074-1088.

Faris, W. (1973), Essential self-adjointness of operators in ordered Hilbert space, Commun. Math. Phys. 30, 23-34.

Faris, W. and R. Lavine (1974), Commutators and self-adjointness of Hamiltonian operators, Commun. Math. Phys. 35, 39-48.

Friedman, C.N. (1972), Perturbations of the Schrödinger equation by potentials with small support, J. Functional Anal. 10, 346-360.

Glazman, I.M. (1965), Direct Methods of Qualitative Spectral Analysis of Singular Differential Operators, Israel Program for Scientific Translations, Jerusalem.

Glimm, J. and A. Jaffe (1969), Singular perturbations of self-adjoint operators, Commun. Pure Appl. Math. 22, 401-414.

Glimm, J. and A. Jaffe (1971), Quantum field theory models, in: Statistical Mechanics and Quantum Field Theory - Les Houches 1970, ed. by C. DeWitt and R. Stora, Gordan and Breach, N.Y.

Glimm, J. and A. Jaffe (1972a), Boson quantum field models, in: Mathematics of Contemporary Physics, ed. by R. Streater, Academic Press, New York.

Glimm, J. and A. Jaffe (1972b), The $\lambda\phi_2^4$ quantum field theory without cutoffs, IV. Perturbations of the Hamiltonian, J. Mathematical Phys. 13, 1568-1584.

Goldstein, J.A. (1972), Some counterexamples involving self-adjoint operators, Rocky Mtn. J. Math. 2, 143-149.

Gross, L. (1972), Existence and uniqueness of physical ground states, J. Functional Anal. 10, 52-108.

Gross, L. ( ), Logarithmic Sobolev inequalities, Amer. J. Math.

Gross, L. ( ), Analytic vectors for representations of the canonical commutation relations and non-degeneracy of ground states, J. Functional Anal..

Goodman, R. (1969), Analytic domination by fractional powers of a positive operator, J. Functional Anal. 3, 246-264.

Gustafson, K. and P. Rejto (1973), Some essentially self-adjoint Dirac operators with spherically symmetric potentials, Israel J. Math. 14, 63-75.

Hellwig, G. (1967), Differential Operators of Mathematical Physics, Addison-Wesley, Reading, Mass..

Hoegh-Krohn, R. (1971), A general class of quantum fields without cutoff in two space-time dimensions, Commun. Math. Phys. 21, 244-255.

Jörgens, K. and J. Weidmann (1973), Spectral Properties of Hamiltonian Operators, Springer, Berlin.

Kalf, H. (1973), Self-adjointness for strongly singular potentials with a $-|x|^2$ fall-off at infinity, Math. Z. 133, 249-255.

Kalf, H. and J. Walter (1972), Strongly singular potentials and essential self-adjointness of singular elliptic operators in $C_o^\infty(R^n \setminus \{0\})$, J. Functional Anal. 10, 114-130.

Kalf, H. and J. Walter (1973), Note on a paper of Simon on essentially self-adjoint Schrödinger operators with singular potentials, Arch. Rat. Mech. Anal. 52, 258-260.

Kato, T. (1966), Perturbation Theory for Linear Operators, Springer, New York.

Kato, T. (1972), Schrödinger operators with singular potentials, Israel J. Math. 13, 135-148.

Kato, T. (1973), A remark to the preceding paper by Chernoff, J. Functional Anal. 12, 415-417.

Kato, T. (1974), A second look at the essential self-adjointness of the Schrödinger operators, in Physical Reality and Mathematical Description, edited by C.P. Enz and J. Mehra, Reidel, Dordrecht, Holland.

Konrady, J. (1971), Almost positive perturbations of positive self-adjoint operators, Commun. Math. Phys. 22, 295-300.

Krein, M.G. (1947), The theory of self-adjoint extensions of semi-bounded Hermitian operators and its applications I, Math. Sbornik 20, 431-495, II, Math. Sbornik 21, 365-404 (Russian).

Masson, D. and W.K. McClary (1972), Classes of C^∞ vectors and essential self-adjointness, J. Functional Anal. 10, 19-32.

McBryan, O.A. (1973), Local generators for the Lorentz group in the $P(\phi)_2$ model, Nuovo Cimento 18A, 654-662.

McBryan, O.A. ( ), Self-adjointness of relatively bounded quadratic forms and operators, preprint Dept. of Math., Univ. of Toronto.

McIntosh, A. (1970a), Bilinear forms in Hilbert space, J. Math. Mech. 19, 1027-1045.

McIntosh, A. (1970b), Hermitian bilinear forms which are not semi-bounded, Bull. Amer. Math. Soc. 76, 732-737.

McIntosh, A. (1972), On the comparability of $A^{\frac{1}{2}}$ and $A^{*\frac{1}{2}}$, Proc. Amer. Math. Soc. 32, 430-434.

Nelson, E. (1959), Analytic vectors, Ann. Math. 70, 572-615.

Nelson, E. (1964), Feynman integrals and the Schrödinger equation, J. Mathematical Phys. 5, 332-343.

Nelson, E. (1966), A quartic interaction in two dimensions, in: Mathematical Theory of Elementary Particles, ed. by R. Goodman and I.E. Segal, M.I.T. Press, Cambridge.

Nelson, E. (1969), Topics in Dynamics I : Flows, Princeton Univ. Press, Princeton, N.J..

Nelson, E. (1972), Time-ordered operator products of sharp-time quadratic forms, J. Functional Anal. 11, 211-219.

Nelson, E. (1973), The free Markoff field, J. Functional Anal. 12, 211-227.

Nelson, E. and W.F. Stinespring (1959), Representation of elliptic operators in an enveloping algebra, Amer. J. Math. 81, 547-560.

Nussbaum, A.E. (1965), Quasi-analytic vectors, Arkiv för Matematik 6, 179-191.

Nussbaum, A.E. (1969), A note on quasi-analytic vectors, Studia Math. 33, 305-309.

Okazawa, N. (1971), A perturbation theorem for linear contraction semigroups on reflexive Banach spaces, Proc. Japan Acad. 47, suppl. II, 947-949.

Okazawa, N. (1973), Perturbations of linear m-accretive operators, Proc. Amer. Math. Soc. 37, 169-174.

Parrott, S. (1969), Uniqueness of the Hamiltonian in quantum field theories, Commun. Math. Phys. 13, 68-72.

Protter, M.H. and H.F. Weinberger (1967), Maximum Principles in Differential Equations, Prentice Hall, Englewood Cliffs.

Radin, C. ( ), Some remarks on the evolution of a Schrödinger particle in an attractive $1/r^2$ potential, preprint, Dept. of Math., Rockefeller Univ., New York.

Rosen, L. (1970), A $\lambda\phi^{2n}$ field theory without cutoffs, Commun. Math. Phys. 16, 157-183.

Schechter, M. (1971), Spectra of Partial Differential Operators, North-Holland, Amsterdam.

Schechter, M. (1972), Hamiltonians for singular potentials, Indiana Univ. Math. J. 22, 483-503.

Schmincke, U.-W. (1972), Essential self-adjointness of a Schrödinger operator with strongly singular potential, Math. Z. 124, 47-50.

Schmincke, U.-W. (1973), A spectral gap theorem for Dirac operators with central field, Math. Z. 131, 351-356.

Schonbek, T.P. (1973), Notes to a paper by C.N. Friedman, J. Functional Anal. 14, 281-294.

Schwartz, L. (1966), Théorie des Distributions, Hermann, Paris.

Segal, I. (1970), Construction of non-linear local quantum processes: I, Ann. Math. 92, 462-481.

Segal, I. (1971), Construction of non-linear local quantum processes: II, Invent. Math. 14, 211-241.

Semenov, Yu.A. (1972), On the Lie-Trotter theorems in L_p-spaces, preprint, Inst. Theoretical Phys., Acad. Sci. Ukrainian SSR, Kiev.

Simon, B. (1971a), Quantum Mechanics for Hamiltonians Defined as Quadratic Forms, Princeton Univ. Press, Princeton, N.J..

Simon, B. (1971b), The theory of semi-analytic vectors: A new proof of a theorem of Masson and McClary, Indiana Univ. Math. J. 20, 1145-1151.

Simon, B. (1971c), Determination of eigenvalues by divergent perturbation series, Adv. Math. 7, 240-253.

Simon, B. (1972), Topics in functional analysis, in Mathematics of Contemporary Physics, edited by R.F. Streater, Academic Press, London.

Simon, B. (1973a), Essential self-adjointness of Schrödinger operators with positive potentials, Math. Ann. 201, 211-220.

Simon, B. (1973b), Essential self-adjointness of Schrödinger operators with singular potentials, Arch. Rat. Mech. Anal. 52, 44-48.

Simon, B. (1973c), Schrödinger operators with singular magnetic vector potentials, Math. Z. 131, 361-370.

Simon, B. (1973d), Quadratic forms and Klauder's phenomenon: A remark on very singular perturbations, J. Functional Anal. 14, 295-298.

Simon, B. (1973e), Ergodic semigroups of positivity preserving self-adjoint operators, J. Functional Anal. 12, 335-339.

Simon, B. and R. Hoegh-Krohn (1972), Hypercontractive semigroups and two dimensional self-coupled Bose fields, J. Funct. Anal. 9, 121-180.

Sloan, A.D. ( ), Analytic domination with quadratic form type estimates and non-degeneracy of ground states in quantum field theory, Trans. A.M.S..

Sloan, A.D. ( ), Analytic domination by fractional powers with linear estimates, preprint, Math.Dept., Georgia Inst. Tech., Atlanta

Stein, E.M. (1970), Singular Integrals and Differentiability Properties of Functions, Princeton Univ. Press, Princeton, N.J..

Stein, E.M. and G. Weiss (1971), Introduction to Fourier Analysis on Euclidean Spaces, Princeton Univ. Press, Princeton, N.J..

Velo, G. and A. Wightman (eds.) (1973), Constructive Quantum Field Theory, Springer, Berlin.

Wüst, R. (1971), Generalizations of Rellich's theorem on perturbation of (essentially) self-adjoint operators, Math. Z. 119, 276-280.

Wüst, R. (1972), Holomorphic operator families and stability of self-adjointness, Math. Z. 125, 349-358.

Wüst, R. (1973), A convergence theorem for self-adjoint operators applicable to Dirac operators with cutoff potentials, Math. Z. 131, 339-349.

Yakimov, Ya.M. (1974), On a criterion of essential self-adjointness of operators, preprint, Inst. Theoretical Phys., Acad. Sci. Ukrainian SSR, Kiev.

ol. 277: Séminaire Banach. Edité par C. Houzel. VII, 229 pages. 972. DM 20,-

ol. 278: H. Jacquet, Automorphic Forms on GL(2). Part II. XIII, 142 ages. 1972. DM 16,-

ol. 279: R. Bott, S. Gitler and I. M. James, Lectures on Algebraic and ifferential Topology. V, 174 pages. 1972. DM 18,-

ol. 280: Conference on the Theory of Ordinary and Partial Differenal Equations. Edited by W. N. Everitt and B. D. Sleeman. XV, 367 ages. 1972. DM 26,-

ol. 281: Coherence in Categories. Edited by S. Mac Lane. VII, 235 ages. 1972. DM 20,-

ol. 282: W. Klingenberg und P. Flaschel, Riemannsche Hilbertmanigfaltigkeiten. Periodische Geodätische. VII, 211 Seiten. 1972. M 20,-

ol. 283: L. Illusie, Complexe Cotangent et Déformations II. VII, 304 ages. 1972. DM 24,-

ol. 284: P. A. Meyer, Martingales and Stochastic Integrals I. VI, 89 ages. 1972. DM 16,-

ol. 285: P. de la Harpe, Classical Banach-Lie Algebras and Banachie Groups of Operators in Hilbert Space. III, 160 pages. 1972. M 16,-

ol. 286: S. Murakami, On Automorphisms of Siegel Domains. V, 5 pages. 1972. DM 16,-

ol. 287: Hyperfunctions and Pseudo-Differential Equations. Edited y H. Komatsu. VII, 529 pages. 1973. DM 36,-

ol. 288: Groupes de Monodromie en Géométrie Algébrique. (SGA 7 . Dirigé par A. Grothendieck. IX, 523 pages. 1972. DM 50,-

ol. 289: B. Fuglede, Finely Harmonic Functions. III, 188. 1972. M 18,-

ol. 290: D. B. Zagier, Equivariant Pontrjagin Classes and Applicaons to Orbit Spaces. IX, 130 pages. 1972. DM 16,-

ol. 291: P. Orlik, Seifert Manifolds. VIII, 155 pages. 1972. DM 16,-

ol. 292: W. D. Wallis, A. P. Street and J. S. Wallis, Combinatorics: oom Squares, Sum-Free Sets, Hadamard Matrices. V, 508 pages. 972. DM 50,-

ol. 293: R. A. DeVore, The Approximation of Continuous Functions y Positive Linear Operators. VIII, 289 pages. 1972. DM 24,-

ol. 294: Stability of Stochastic Dynamical Systems. Edited by R. F. urtain. IX, 332 pages. 1972. DM 26,-

ol. 295: C. Dellacherie, Ensembles Analytiques, Capacités, Mesures e Hausdorff. XII, 123 pages. 1972. DM 16,-

ol. 296: Probability and Information Theory II. Edited by M. Behara, . Krickeberg and J. Wolfowitz. V, 223 pages. 1973. DM 20,-

ol. 297: J. Garnett, Analytic Capacity and Measure. IV, 138 pages. 972. DM 16,-

ol. 298: Proceedings of the Second Conference on Compact Transormation Groups. Part 1. XIII, 453 pages. 1972. DM 32,-

ol. 299: Proceedings of the Second Conference on Compact Transormation Groups. Part 2. XIV, 327 pages. 1972. DM 26,-

ol. 300: P. Eymard, Moyennes Invariantes et Représentations Uniaires. II. 113 pages. 1972. DM 16,-

ol. 301: F. Pittnauer, Vorlesungen über asymptotische Reihen. VI, 86 Seiten. 1972. DM 18,-

ol. 302: M. Demazure, Lectures on p-Divisible Groups. V, 98 pages. 972. DM 16,-

ol. 303: Graph Theory and Applications. Edited by Y. Alavi, D. R. Lick nd A. T. White. IX, 329 pages. 1972. DM 26,-

ol. 304: A. K. Bousfield and D. M. Kan, Homotopy Limits, Compleons and Localizations. V, 348 pages. 1972. DM 26,-

ol. 305: Théorie des Topos et Cohomologie Etale des Schémas. ome 3. (SGA 4). Dirigé par M. Artin, A. Grothendieck et J. L. Verdier. I, 640 pages. 1973. DM 50,-

ol. 306: H. Luckhardt, Extensional Gödel Functional Interpretation. I, 161 pages. 1973. DM 18,-

ol. 307: J. L. Bretagnolle, S. D. Chatterji et P.-A. Meyer, Ecole d'été e Probabilités: Processus Stochastiques. VI, 198 pages. 1973. M 20,-

ol. 308: D. Knutson, λ-Rings and the Representation Theory of the ymmetric Group. IV, 203 pages. 1973. DM 20,-

Vol. 309: D. H. Sattinger, Topics in Stability and Bifurcation Theory. VI, 190 pages. 1973. DM 18,-

Vol. 310: B. Iversen, Generic Local Structure of the Morphisms in Commutative Algebra. IV, 108 pages. 1973. DM 16,-

Vol. 311: Conference on Commutative Algebra. Edited by J. W. Brewer and E. A. Rutter. VII, 251 pages. 1973. DM 22,-

Vol. 312: Symposium on Ordinary Differential Equations. Edited by W. A. Harris, Jr. and Y. Sibuya. VIII, 204 pages. 1973. DM 22,-

Vol. 313: K. Jörgens and J. Weidmann, Spectral Properties of Hamiltonian Operators. III, 140 pages. 1973. DM 16,-

Vol. 314: M. Deuring, Lectures on the Theory of Algebraic Functions of One Variable. VI, 151 pages. 1973. DM 16,-

Vol. 315: K. Bichteler, Integration Theory (with Special Attention to Vector Measures). VI, 357 pages. 1973. DM 26,-

Vol. 316: Symposium on Non-Well-Posed Problems and Logarithmic Convexity. Edited by R. J. Knops. V, 176 pages. 1973. DM 18,-

Vol. 317: Séminaire Bourbaki – vol. 1971/72. Exposés 400–417. IV, 361 pages. 1973. DM 26,-

Vol. 318: Recent Advances in Topological Dynamics. Edited by A. Beck, VIII, 285 pages. 1973. DM 24,-

Vol. 319: Conference on Group Theory. Edited by R. W. Gatterdam and K. W. Weston. V, 188 pages. 1973. DM 18,-

Vol. 320: Modular Functions of One Variable I. Edited by W. Kuyk. V, 195 pages. 1973. DM 18,-

Vol. 321: Séminaire de Probabilités VII. Edité par P. A. Meyer. VI, 322 pages. 1973. DM 26,-

Vol. 322: Nonlinear Problems in the Physical Sciences and Biology. Edited by I. Stakgold, D. D. Joseph and D. H. Sattinger. VIII, 357 pages. 1973. DM 26,-

Vol. 323: J. L. Lions, Perturbations Singulières dans les Problèmes aux Limites et en Contrôle Optimal. XII, 645 pages. 1973. DM 42,-

Vol. 324: K. Kreith, Oscillation Theory. VI, 109 pages. 1973. DM 16,-

Vol. 325: Ch.-Ch. Chou, La Transformation de Fourier Complexe et L'Equation de Convolution. IX, 137 pages. 1973. DM 16,-

Vol. 326: A. Robert, Elliptic Curves. VIII, 264 pages. 1973. DM 22,-

Vol. 327: E. Matlis, 1-Dimensional Cohen-Macaulay Rings. XII, 157 pages. 1973. DM 18,-

Vol. 328: J. R. Büchi and D. Siefkes, The Monadic Second Order Theory of All Countable Ordinals. VI, 217 pages. 1973. DM 20,-

Vol. 329: W. Trebels, Multipliers for (C, α)-Bounded Fourier Expansions in Banach Spaces and Approximation Theory. VII, 103 pages. 1973. DM 16,-

Vol. 330: Proceedings of the Second Japan-USSR Symposium on Probability Theory. Edited by G. Maruyama and Yu. V. Prokhorov. VI, 550 pages. 1973. DM 36,-

Vol. 331: Summer School on Topological Vector Spaces. Edited by L. Waelbroeck. VI, 226 pages. 1973. DM 20,-

Vol. 332: Séminaire Pierre Lelong (Analyse) Année 1971-1972. V, 131 pages. 1973. DM 16,-

Vol. 333: Numerische, insbesondere approximationstheoretische Behandlung von Funktionalgleichungen. Herausgegeben von R. Ansorge und W. Törnig. VI, 296 Seiten. 1973. DM 24,-

Vol. 334: F. Schweiger, The Metrical Theory of Jacobi-Perron Algorithm. V, 111 pages. 1973. DM 16,-

Vol. 335: H. Huck, R. Roitzsch, U. Simon, W. Vortisch, R. Walden, B. Wegner und W. Wendland, Beweismethoden der Differentialgeometrie im Großen. IX, 159 Seiten. 1973. DM 18,-

Vol. 336: L'Analyse Harmonique dans le Domaine Complexe. Edité par E. J. Akutowicz. VIII, 169 pages. 1973. DM 18,-

Vol. 337: Cambridge Summer School in Mathematical Logic. Edited by A. R. D. Mathias and H. Rogers. IX, 660 pages. 1973. DM 42,-

Vol: 338: J. Lindenstrauss and L. Tzafriri, Classical Banach Spaces. IX, 243 pages. 1973. DM 22,-

Vol. 339: G. Kempf, F. Knudsen, D. Mumford and B. Saint-Donat, Toroidal Embeddings I. VIII, 209 pages. 1973. DM 20,-

Vol. 340: Groupes de Monodromie en Géométrie Algébrique. (SGA 7 II). Par P. Deligne et N. Katz. X, 438 pages. 1973. DM 40,-

Vol. 341: Algebraic K-Theory I, Higher K-Theories. Edited by H. Bass. XV, 335 pages. 1973. DM 26,-

Vol. 342: Algebraic K-Theory II, "Classical" Algebraic K-Theory, and Connections with Arithmetic. Edited by H. Bass. XV, 527 pages. 1973. DM 36,-

Vol. 343: Algebraic K-Theory III, Hermitian K-Theory and Geometric Applications. Edited by H. Bass. XV, 572 pages. 1973. DM 38,-

Vol. 344: A. S. Troelstra (Editor), Metamathematical Investigation of Intuitionistic Arithmetic and Analysis. XVII, 485 pages. 1973. DM 34,-

Vol. 345: Proceedings of a Conference on Operator Theory. Edited by P. A. Fillmore. VI, 228 pages. 1973. DM 20,-

Vol. 346: Fučík et al., Spectral Analysis of Nonlinear Operators. II, 287 pages. 1973. DM 26,-

Vol. 347: J. M. Boardman and R. M. Vogt, Homotopy Invariant Algebraic Structures on Topological Spaces. X, 257 pages. 1973. DM 22,-

Vol. 348: A. M. Mathai and R. K. Saxena, Generalized Hypergeometric Functions with Applications in Statistics and Physical Sciences. VII, 314 pages. 1973. DM 26,-

Vol. 349: Modular Functions of One Variable II. Edited by W. Kuyk and P. Deligne. V, 598 pages. 1973. DM 38,-

Vol. 350: Modular Functions of One Variable III. Edited by W. Kuyk and J.-P. Serre. V, 350 pages. 1973. DM 26,-

Vol. 351: H. Tachikawa, Quasi-Frobenius Rings and Generalizations. XI, 172 pages. 1973. DM 18,-

Vol. 352: J. D. Fay, Theta Functions on Riemann Surfaces. V, 137 pages. 1973. DM 16,-

Vol. 353: Proceedings of the Conference. on Orders, Group Rings and Related Topics. Organized by J. S. Hsia, M. L. Madan and T. G. Ralley. X, 224 pages. 1973. DM 20,-

Vol. 354: K. J. Devlin, Aspects of Constructibility. XII, 240 pages. 1973. DM 22,-

Vol. 355: M. Sion, A Theory of Semigroup Valued Measures. V, 140 pages. 1973. DM 16,-

Vol. 356: W. L. J. van der Kallen, Infinitesimally Central-Extensions of Chevalley Groups. VII, 147 pages. 1973. DM 16,-

Vol. 357: W. Borho, P. Gabriel und R. Rentschler, Primideale in Einhüllenden auflösbarer Lie-Algebren. V, 182 Seiten. 1973. DM 18,-

Vol. 358: F. L. Williams, Tensor Products of Principal Series Representations. VI, 132 pages. 1973. DM 16,-

Vol. 359: U. Stammbach, Homology in Group Theory. VIII, 183 pages. 1973. DM 18,-

Vol. 360: W. J. Padgett and R. L. Taylor, Laws of Large Numbers for Normed Linear Spaces and Certain Fréchet Spaces. VI, 111 pages. 1973. DM 16,-

Vol. 361: J. W. Schutz, Foundations of Special Relativity: Kinematic Axioms for Minkowski Space Time. XX, 314 pages. 1973. DM 26,-

Vol. 362: Proceedings of the Conference on Numerical Solution of Ordinary Differential Equations. Edited by D. Bettis. VIII, 490 pages. 1974. DM 34,-

Vol. 363: Conference on the Numerical Solution of Differential Equations. Edited by G. A. Watson. IX, 221 pages. 1974. DM 20,-

Vol. 364: Proceedings on Infinite Dimensional Holomorphy. Edited by T. L. Hayden and T. J. Suffridge. VII, 212 pages. 1974. DM 20,-

Vol. 365: R. P. Gilbert, Constructive Methods for Elliptic Equations. VII, 397 pages. 1974. DM 26,-

Vol. 366: R. Steinberg, Conjugacy Classes in Algebraic Groups (Notes by V. V. Deodhar). VI, 159 pages. 1974. DM 18,-

Vol. 367: K. Langmann und W. Lütkebohmert, Cousinverteilungen und Fortsetzungssätze. VI, 151 Seiten. 1974. DM 16,-

Vol. 368: R. J. Milgram, Unstable Homotopy from the Stable Point of View. V, 109 pages. 1974. DM 16,-

Vol. 369: Victoria Symposium on Nonstandard Analysis. Edited by A. Hurd and P. Loeb. XVIII, 339 pages. 1974. DM 26,-

Vol. 370: B. Mazur and W. Messing, Universal Extensions an One Dimensional Crystalline Cohomology. VII, 134 pages. 1974 DM 16,-

Vol. 371: V. Poenaru, Analyse Différentielle. V, 228 pages. 1974 DM 20,-

Vol. 372: Proceedings of the Second International Conferenc on the Theory of Groups 1973. Edited by M. F. Newman. VI 740 pages. 1974. DM 48,-

Vol. 373: A. E. R. Woodcock and T. Poston, A Geometrical Stud of the Elementary Catastrophes. V, 257 pages. 1974. DM 22,

Vol. 374 S. Yamamuro, Differential Calculus in Topologica Linear Spaces. IV, 179 pages. 1974. DM 18,-

Vol. 375: Topology Conference 1973. Edited by R. F. Dickma Jr. and P. Fletcher. X, 283 pages. 1974. DM 24,-

Vol. 376: D. B. Osteyee and I. J. Good, Information, Weight o Evidence, the Singularity between Probability Measures an Signal Detection. XI, 156 pages. 1974. DM 16.-

Vol. 377: A. M. Fink, Almost Periodic Differential Equation VIII, 336 pages. 1974. DM 26,-

Vol. 378: TOPO 72 – General Topology and its Application Proceedings 1972. Edited by R. Alò, R. W. Heath and J. Nagat XIV, 651 pages. 1974. DM 50,-

Vol. 379: A. Badrikian et S. Chevet, Mesures Cylindrique Espaces de Wiener et Fonctions Aléatoires Gaussiennes. X 383 pages. 1974. DM 32,-

Vol. 380: M. Petrich, Rings- and Semigroups. VIII, 182 page 1974. DM 18,-

Vol. 381: Séminaire de Probabilités VIII. Edité par P. A. Meye IX, 354 pages. 1974. DM 32,-

Vol. 382: J. H. van Lint, Combinatorial Theory Seminar Ein hoven University of Technology. VI, 131 pages. 1974. DM 18,

Vol. 383: Séminaire Bourbaki – vol. 1972/73. Exposés 418-43 IV, 334 pages. 1974. DM 30,-

Vol. 384: Functional Analysis and Applications, Proceeding 1972. Edited by L. Nachbin. V, 270 pages. 1974. DM 22,-

Vol. 385: J. Douglas Jr. and T. Dupont, Collocation Methods f Parabolic Equations in a Single Space Variable (Based on C Piecewise-Polynomial Spaces). V, 147 pages. 1974. DM 16,

Vol. 386: J. Tits, Buildings of Spherical Type and Finite B Pairs. IX, 299 pages. 1974. DM 24,-

Vol. 387: C. P. Bruter, Eléments de la Théorie des Matroïde V, 138 pages. 1974. DM 18,-

Vol. 388: R. L. Lipsman, Group Representations. X, 166 page 1974. DM 20,-

Vol. 389: M.-A. Knus et M. Ojanguren, Théorie de la Descen et Algèbres d' Azumaya. IV, 163 pages. 1974. DM 20,-

Vol. 390: P. A. Meyer, P. Priouret et F. Spitzer, Ecole d'Eté c Probabilités de Saint-Flour III – 1973. Edité par A. Badriki et P.-L. Hennequin. VIII, 189 pages. 1974. DM 20,-

Vol. 391: J. Gray, Formal Category Theory: Adjointness for Categories. XII, 282 pages. 1974. DM 24,-

Vol. 392: Géométrie Différentielle, Colloque, Santiago Compostela, Espagne 1972. Edité par E. Vidal. VI, 225 page 1974. DM 20,-

Vol. 393: G. Wassermann, Stability of Unfoldings. IX, 164 page 1974. DM 20,-

Vol. 394: W. M. Patterson 3rd, Iterative Methods for the Soluti of a Linear Operator Equation in Hilbert Space – A Surve III, 183 pages. 1974. DM 20,-

Vol. 395: Numerische Behandlung nichtlinearer Integrodiffere tial- und Differentialgleichungen. Tagung 1973. Herausgegeb von R. Ansorge und W. Törnig. VII, 313 Seiten. 1974. DM 28

Vol. 396: K. H. Hofmann, M. Mislove and A. Stralka, The Pontr agin Duality of Compact O-Dimensional Semilattices and Applications. XVI, 122 pages. 1974. DM 18,-

Vol. 397: T. Yamada, The Schur Subgroup of the Brauer Grou V, 159 pages. 1974. DM 18,-

Vol. 398: Théories de l'Information, Actes des Rencontres c Marseille-Luminy, 1973. Edité par J. Kampé de Fériet et C. Pica XII, 201 pages. 1974. DM 23,-